MULTIDIMENSIONAL FILTER BANKS AND WAVELETS
Basic Theory and Cosine Modulated Filter Banks

edited by

Sankar Basu
IBM Thomas J. Watson Research Center
Bernard Levy
University of California, Davis

A Special Issue of
**MULTIDIMENSIONAL SYSTEMS
AND SIGNAL PROCESSING**
An International Journal
Volume 7, Nos. 3/4 (1996)

KLUWER ACADEMIC PUBLISHERS
Boston / Dordrecht / London

MULTIDIMENSIONAL FILTER BANKS AND WAVELETS
Basic Theory and Cosine Modulated Filter Banks

edited by

Sankar Basu
IBM Thomas J. Watson Research Center
Bernard Levy
University of California, Davis

A Special Issue of
MULTIDIMENSIONAL SYSTEMS
AND SIGNAL PROCESSING
An International Journal
Volume 7, Nos. 3/4 (1996)

KLUWER ACADEMIC PUBLISHERS
Boston / Dordrecht / London

MULTIDIMENSIONAL SYSTEMS AND SIGNAL PROCESSING

An International Journal

Volume 7, Nos. 3/4, July/October 1996

Special Issue: Multidimensional Filter Banks and Wavelets: Basic Theory and Cosine Modulated Filter Banks
Guest Editors: Sankar Basu and Bernard Levy

Editorial...*Nirmal K. Bose*

Introduction..

Theory and Design of Two-Dimensional Filter Banks: A Review............. ....*Yuan-Pei Lin and P. P. Vaidyanathan*

Local Orthogonal Bases I: Construction ....*R. Bernardini and J. Kovačević*

Local Orthogonal Bases II: Window Design... ....*R. Bernardini and J. Kovačević*

Contributing Authors...

Distributors for North America:
Kluwer Academic Publishers
101 Philip Drive
Assinippi Park
Norwell, Massachusetts 02061 USA

Distributors for all other countries:
Kluwer Academic Publishers Group
Distribution Centre
Post Office Box 322
3300 AH Dordrecht, THE NETHERLANDS

ISBN 978-1-4419-5163-2

Library of Congress Cataloging-in-Publication Data

A C.I.P. Catalogue record for this book is available
from the Library of Congress.

Printed on acid-free paper.

Printed in the United States of America

Multidimensional Systems and Signal Processing, 7, 257 (1996)

Editorial

This issue on the topic of wavelets and multiresolution signal processing is the fourth special issue sponsored by this journal. There were an abundance of good papers and when the guest editors, Dr. S. Basu from IBM, Yorktown Heights, and Professor B. Levy from the University of California, Davis, submitted their table of contents for accepted papers after careful review, the number of pages originally planned for allocation had already been exceeded by an appreciable margin. Viewed from an optimistic standpoint, this problem was interpreted to be due to an expanding base of interest in the subject-matter with profound implications in future research. So, instead of trimming further as necessitated by page counts, an acceptable solution was devised with the cooperation of Mr. Robert Holland of Kluwer Academic Publishers. A set of papers were included in this joint issue of Volume 7 including a detailed review that will hopefully attract the attention of specialists as well as non-specialists who might want to join the bandwagon of contributors in the future. The remaining papers, which have a blend of theory as well as selected applications like image coding, tomography, and computer graphics are included in the first joint issue of Volume 8, which is to follow shortly. Both these joint issues are planned for separate printing as hardcover volumes.

The guest editors spared no efforts in pursuing a fair but comprehensive and constructive reviewing scheme consistent with the policies of this journal. About fifty expert reviewers helped the guest editors in arriving at their final selections for this special issue. Some of the papers that could not be included in the special issue are likely to appear at a later date in one or more of the regular issues after final processing and acceptance.

We hope that the readers will be stimulated by the proven and potential possibilities of some of the results ingrained in multidimensional systems theory for adaptation and use in wavelet theory and multiresolution signal processing, an area of research which is itself a fertile arena for further theoretical developments and an unusually broad spectrum of applications. For instance, the relevance in multidimensional systems of Groebner bases from polynomial ideal theory and their construction by algorithms devised by Buchberger and others were noted over a decade back (see, for example, N. K. Bose, *Multidimensional Systems: Progress, Directions, and Open Problems*; D. Reidel Publishing Co., Dordrecht, Holland, 1985) and their potentials in the topic of concern here are yet to be fully realized. I thank the guest editors for their efforts in bringing out this timely special issue, where the benefits to be derived from interactions between researchers in a more mature discipline on one hand and a more current discipline on the other are underscored.

N. K. Bose
Editor-in-Chief

The following papers will appear in issue 8:1/2 of MULTIDIMENSIONAL SYSTEMS AND SIGNAL PROCESSING. Together with the papers in this issue, this body of work represents a total of four issues developed by Guest Editors Sankar Basu and Bernard Levy. We are grateful for their contributions to MULTIDIMENSIONAL SYSTEMS AND SIGNAL PROCESSING.

Hyungju Park/Ton Kalker/Martin Vetterli - *Gröbner Bases and Multidimensional FIR Multirate Systems*

Charles A. Micchelli/Yuesheng Xu - *Reconstruction and Decomposition Algorithms for Biorthogonal Multiwavelets*

T. G. Marshall, Jr. - *Zero-Phase Filter Bank and Wavelet Coder Matrices: Properties, Triangular Decompositions, and a Fast Algorithm*

Steven A. Benno/José M. F. Moura - *On Translation Invariant Subspaces and Critically Sampled Wavelet Transforms*

Michael Lightstone/Eric Marjani/Sanjit K. Mitra - *Low Bit-Rate Design Considerations for Wavelet-Based Image Coding*

G. Calvagno/R. Rinaldo - *Multiresolution Vector Quantization for Video Coding*

Eric L. Miller/Alan S. Willsky - *Multiscale, Statistical Anomaly Detection Analysis and Algorithms for Linearized Inverse Scattering Problems*

José Fridman/Elias S. Manolakos - *On the Scalability of 2-D Discrete Wavelet Transform Algorithms*

SHORT PAPER

Wayne Lawton - *A Fast Algorithm to Map Functions Forward*

Multidimensional Systems and Signal Processing, 7, 259–262 (1996)
© 1996 Kluwer Academic Publishers, Boston. Manufactured in The Netherlands.

Introduction

The topic of wavelets has gained considerable importance over the last several years. It has proven to be an elegant tool for solving problems in a wide variety of areas of both theoretical and applied endeavour. While certain commonalities in ideas and techniques used in apparently disparate disciplines is responsible for this explosion of activity in the field of wavelets and multiresolution analysis, the topic in turn has also brought together researchers from research communities of diverse variety with further exciting possibilities for the future. Several texts, monographs, edited books and special issues of journals have been published recently on the topic and we are certain that many more will appear in years to come.

The field of multidimensional systems and signal processing, on the other hand, has attained a certain level of maturity over the last two decades through publications in archival literature in the form of several texts, edited books and special issues of journals. We have recently noticed a synergy between the two fields of wavelets and multidimensional systems arising from the fact that many ideas crucial to the wavelet theory are inherently system theoretic in nature. Techniques germane to multidimensional system theory are beginning to find use in multidimensional wavelet design.

While there are many examples of this synergy manifested in the articles included in this two part special issue, and much more still remains to be explored in the future, we may mention only a few. The construction of orthogonal wavelets can be essentially viewed as a circuit and system theoretic problem of design of energy dissipative (passive) filters, the multidimensional version of which has very close ties with the classic multidimensional problem of lumped-distributed passive network synthesis. Recent work on multidimensional wavelet design has benefitted substantially from this body of work on lumped-distributed network systems originally attributed to Youla and collaborators. Groebner basis techniques, matrix completion problems over rings of polynomials or rings of stable rational functions are still other examples, which remain to be completely explored.

Evidence supporting the relevance of the multidimensional version of wavelets in a broad range of topics in science, mathematics and engineering is overwhelming. The wavelet methods have been claimed to impact a wide range of disciplines such as in the engineering areas of signal processing communication theory, computer vision, computer graphics and biomedical applications. Physical problems of turbulent flow, the study of distant galaxies, and of course problems of mathematics, including numerical analysis and statistics have also found applications of wavelet theory. Indeed, a cursory examination of the topics mentioned above would reveal that a large majority of the problems are truely multidimensional in nature.

The present special issue, by contrast, is more focussed. Besides reporting new fundamental techniques and results specific to the multidimensional situation, we hope to be to report the present state of the art in design, implementation of multirate multidimensional filter banks and wavelets, and some of their promising applications in image coding, tomography, radar/sonar detection and computer graphics. In spite of the profusion of literature in the field of wavelets, however, there seems to be a lack of definitive publications that bring into focus the state of the art, the techniques currently in use, and future trends in the field of multidimensional filter banks and wavelets. The present special issue is an attempt to at least partially satisfy this need.

In selecting articles for this special issue we have purposely attempted to adhere to articles that deal with schemes that are inherently nonseparable in the terminology of multidimensional system theory. Thus, role of techniques relying on schemes that are otherwise known as the tensor product schemes or row-column schemes are purposely underplayed. While these schemes may be useful in some practical problems, our purpose is to demonstrate more non-trivial extensions of 1D methods in the multidimensional situation.

Since the present special issue was originally planned, an overwhelmingly large number of excellent articles were received, all of which could not be included simply because of limitations of space. In view of the high quality of these articles and current importance of the topic, a two-fold approach to this difficulty has been adopted. First, the special issue has been split into two parts to satisfy the editorially imposed restrictions on the allowed number of pages in one volume. Second, articles that still could not be accomodated, will be relegated to subsequent regular issues of the journal.

The first part starts with an extensive review of the field of multidimensional multirate filter bank and its relevance to wavelet theory by Lin and Vaidyanathan. The emphasis here is more on multirate filter banks as it applies to multidimensional wavelets. The role of multidimensional sampling in decimation and interpolation is first briefly reviewed by the authors. Several 2D nonseparable design techniques of recent origin potentially useful in image coding are reviewed here. Multidimensional cosine modulated filter banks are also discussed by following recent results of the authors themselves. Finally, the connection between 2D lossless systems and 2D paraunitary filter bank design is also briefly reviewed as well.

The next two papers, with which we conclude the first part of the present special issue, report new results and are somewhat more theoretical in nature. Cosine modulated filter banks mentioned above in the paper by Lin and Vaidyanathan are known to be closely related to the lapped orthogonal transforms, the 1D version of which is now attributed to the work of Malvar in engineering community and is also known as the local orthogonal bases due to Coifman and Meyer in the mathematics literature. The papers by Bernardini and Kovacevic included in the present special issue deals with multidimensional generalization of these 1D concepts. A brief summary of the two papers follow.

In the first of the two papers, *Local Orthogonal Bases: Construction* the authors essentially undertake the problem of construction of interesting and useful set of orthonormal bases of finite support B_i of vector space ν_X of function defined on a set X. Examples of ν_X are $L^2(R^n)$ in the continuous case, or $\ell_2(Z^n)$ in the discrete case. This is achieved by first decomposing B_i's into disjoint union of C_{ij}'s such that the vector spaces over B_i can be seen as a direct sum of suitable vector spaces over C_{ij}'s. For this, associated with C_{ij} a group of involutions Γ, a weight function s that qualifies as a representation of Γ, and a window w

satisfying certain power complementary relations with respect to Γ are chosen. Once this is achieved, it becomes possible to obtain the desired vector spaces with support C_{ij}'s. The conditions that the group Γ, the weight s and the window w must satisfy together in order for this to produce an orthogonal decomposition of v_X is discussed in the first part of the paper.

The paper entitled *Local Orthogonal Bases: Window Design* by the same authors can be viewed as a follow up of the last mentioned paper, and considers the problem of practical window design by focussing on the 2D situation both in the discrete and in the continuous case. The window design method is demonstrated by considering concrete examples where the specific case of B_i's being translations of a set, say B is considered. Example of a window w with hexagonal support is also worked out. The property of so called polyphase normalization i.e., the requirement that as a result of constant input a filter bank produces nonzero output only from the low frequency channel—as is desirable in image compression—is also incorporated into the issue of optimal window design.

Succinctly stated, mapping the classical PR conditions to involutions and projection operators is the key to the multidimensional generalizations reported in these two papers.

The second part of the present special issue, which is to immediately follow the publication of the present volume will include papers on multiwavelets (by Micchelli and Xu); Goebner basis techniques (by Park, Kalker and Vetterli); ladder based techniques (by Marshall); translation invariant wavelet transforms (by Benno and Moura). These papers are somewhat theoretical in flavour. The rest of the papers will consist of a sampling from how multidimensional wavelet based techniques can be used in applied problems. These include two papers on image coding (by Lightstone, Majani, and Mitra and by Calvagno and Rinaldi); and a paper on the use of wavelets in tomography (by Miller and Willsky). The second volume will also see a paper on computational issues, where scalable techniques for wavelet based computation will be discussed (by Fridman and Manolakos). Finally, the issue will conclude with a short note on a problem on function approximation motivated by applications in computer graphics (by Lawton).

While there are many other issues of concern both from the theoretical and practical standpoints in multidimensional versions of the subband coding and wavelet design problem, due to the limited scope of this special issue and editorially imposed page limitations it was not possible to include important research results from many authors on other topics of related interest. These include nonseparable multidimensional Gabor transforms, new linear phase nonseparable filter bank design, and time varying multidimensional filter banks etc. Some of these issues will hopefully appear in the ensuing regular issues of the present journal.

Acknowledgements

Professor N. K. Bose, Chief editor of this journal, suggested to one of us in the Fall of 1993 that the task of compiling a special issue such as this one would indeed be worthwhile. We wish to thank him for his patience and constant encouragement in spite of the time it took us to accomplish the task. Finally, we would like to express our gratitude to all the reviewer's, without critical comments from each of whom this two part special issue would not have been possible. A list of all those who unselfishly devoted their time follows in random order.

Jose Moura (Carnegie Mellon University)
Ephraim Feig (IBM Research)
P. P. Vaidyanathan (Caltech)
Tsuhan Chen (AT&T Bell Lab)
Shi-Dong Li (U of Maryland)
Roberto Bamberger (U of Washington)
R. Vaillancourt (U of Ottowa)
L. F. Chapparo (U of Pittsburgh)
Raymond Wells (Rice University)
Peter Heller (AWARE Inc.)
Ton Kalker (Phillips, Holland)
C. A. Bernstein (U of Maryland)
Tom Parks (Cornell Univeristy)
Imran Shah (PA Consulting Group)
Prabhakar Rao Chitrappu (Dialogic Corporation)
Thomas R. Gardos (Intel)
Jun Zhang (U of Wisconsin, Milwaulkee)
Jelena Kovacevic (AT&T Bell Lab)
Bruce Suter (Air Force Inst. of Technology)
George Cybenko (Dartmouth)
Amro El-Jaroudi (U of Pittsburgh)
Joseph D. Ward (Texas A & M University)
W. Dahmen (Tech. Hoch. Aachen, Germany)
Ronald DeVore (U of S. Carolina)
Ding Xuan Zhou (U of Alberta)
Yuesheng Xu (U of North Dakota)
Alan Bovik (U of Texas at Austin)
Jeffrey Bloom (UC Davis)
Philip Chou (Xerox, Palo Alto)
Ian Kerfoot (U of Illinois)
Wayne Lawton (Nat. U of Singapore)
Truong Nguyen (U of Wisconsin, Madison)
Christine Podilchuk (AT&T Bell Laboratories)
Ken Sauer (U of Notre Dame)
Stan Sclaroff (Boston Univeristy)
Thomas Sederberg (Brigham Young University)
Wim Sweldens (AT&T Bell Lab)
Demetri Terzopoulos (U of Toronto)
Ahmed Tewfik (U of Minnesota)
Shankar Venkataraman (Optivision)
John Villasenor (UCLA)
Andrew Yagle (U of Michigan)

Multidimensional Systems and Signal Processing, 7, 263–330 (1996)
© 1996 Kluwer Academic Publishers, Boston. Manufactured in The Netherlands.

Theory and Design of Two-Dimensional Filter Banks: A Review

YUAN-PEI LIN
Department of Electrical Engineering, 136-93, California Institute of Technology, Pasadena, CA 91125

P. P. VAIDYANATHAN   ppvnath@sys.caltech.edu
Department of Electrical Engineering, 136-93, California Institute of Technology, Pasadena, CA 91125

Abstract. There has been considerable interest in the design of multidimensional (MD) filter banks. MD filter banks find application in subband coding of images and video data. MD filter banks can be designed by cascading one-dimensional (1D) filter banks in the form of a tree structure. In this case, the individual analysis and synthesis filters are separable and the filter bank is called a separable filter bank. MD filter banks with nonseparable filters offer more flexibility and usually provide better performance. Nonetheless, their design is considerably more difficult than separable filter banks. The purpose of this paper is to provide an overview of developments in this field on the design techniques for MD filter banks, mostly two-dimensional (2D) filter banks. In some image coding applications, the 2D two-channel filter banks are of great importance, particularly the filter bank with diamond-shaped filters. We will present several design techniques for the 2D two-channel nonseparable filter banks. As the design of MD filters are not as tractable as that of 1D filters, we seek design techniques that do not involve direct optimization of MD filters. To facilitate this, transformations that turn a separable MD filter bank into a nonseparable one are developed. Also, transformations of 1D filter banks to MD filter banks are investigated. We will review some designs of MD filter banks using transformations. In the context of 1D filter bank design, the cosine modulated filter bank (CMFB) is well-known for its design and implementation efficiency. All the analysis filters are cosine modulated versions of a prototype filter. The design cost of the filter bank is equivalent to that of the prototype and the implementation complexity is comparable to that of the prototype plus a low-complexity matrix. The success with 1D CMFB motivate the generalization to the 2D case. We will construct the 2D CMFB by following a very close analogy of 1D case. It is well-known that the 1D lossless systems can be characterized by state space description. In 1D, the connection between the losslessness of a transfer matrix and the unitariness of the realization matrix is well-established. We will present the developments on the study of 2D lossless systems. As in 1D case, the 2D FIR lossless systems can be characterized in terms of state space realizations. We will review this, and then address the factorizability of 2D FIR lossless systems by using the properties of state space realizations.

1. Introduction

One-dimensional (1D) filter banks have been shown to be very useful in subband coding applications. The typical applications of 1D filter banks are in the coding of speech and music [55]. For multidimensional (MD) filter banks, applications include coding and compression of images and video data. Consider the filter bank in Fig. 1.1. Two of the basic building blocks are the decimator $\mathbf{M}$ and the expander $\mathbf{M}$. In 1D systems, the M-fold decimator keeps only the samples that are multiples of M. In D-dimensional systems, the decimator $\mathbf{M}$ is a $D \times D$ nonsingular integer matrix. The decimation matrix $\mathbf{M}$ keeps those samples that are on the lattice generated by $\mathbf{M}$. The lattice generated by an integer matrix $\mathbf{M}$ is the set of integer vectors of the form

$\mathbf{Md}$, for some $D \times 1$ integer vector $\mathbf{d}$.

7

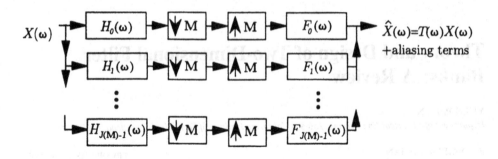

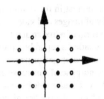

Figure 1.1. $J(\mathbf{M})$-channel maximally decimated filter bank, where $J(\mathbf{M}) = |\det \mathbf{M}|$.

o Integers

• Integers on lattice of $\mathbf{Q}$

Figure 1.2. The lattice generated by $\mathbf{Q}$, the quincunx lattice.

For example, let $\mathbf{M}$ be the quincunx matrix $\mathbf{Q}$ defined as

$$\mathbf{Q} = \begin{pmatrix} 1 & 1 \\ -1 & 1 \end{pmatrix}. \tag{1.1}$$

The lattice of $\mathbf{Q}$ is as shown in Fig. 1.2 and is called the quincunx lattice. The output of the decimator $\mathbf{Q}$ contains only the samples on the lattice of $\mathbf{Q}$. Suppose the system in Fig. 1.1 has input $X(\mathbf{z})$, then the output of the filter bank $\widehat{X}(\mathbf{z})$ consists of $X(\mathbf{z})T(\mathbf{z})$ and some aliasing terms. When the output is free from aliasing error, the system is LTI with transfer function $T(\mathbf{z})$, called the distortion function. If $T(\mathbf{z})$ is a delay, the alias-free filter bank is said to have perfect reconstruction.

The simplest way to design MD filter banks is to cascade 1D filter banks in the form of a tree structure. In these tree-structured filter banks, the decimation matrix $\mathbf{M}$ is diagonal and data is processed in each dimension separately. This type of systems is referred to as separable. For a D dimensional separable filter bank, design cost is equivalent to D times that of 1D filter banks; complexity of design and implementation grows linearly with the number of dimensions. However, in this case, the supports of the analysis and synthesis filter are D-dimensional rectangles, e.g., rectangles in two-dimensional (2D) case. In nonseparable filter banks, the supports of the analysis and synthesis filters could

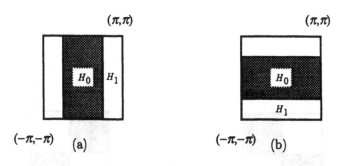

Figure 1.3. Two types of support configurations for separable two-channel filter banks.

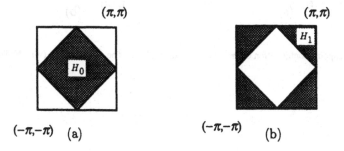

Figure 1.4. The support configuration of the diamond filter bank. (a) The lowpass analysis and synthesis filters. (b) The highpass analysis and synthesis filters.

have a variety of shapes, e.g., parallelepipeds that are not rectangles. Usually different support configurations are desired for different applications. For example, consider 2D two-channel systems. If we use a separable system, the support configuration will be as in Fig. 1.3(a) or Fig. 1.3(b) when the analysis and synthesis filters have real coefficients. (The analysis and the synthesis filters typically have the same supports; only the supports of the analysis filters are shown.) Because the human eye is less sensitive to high frequency component, in some applications of image coding it is desired that one subband has as much low frequency information as possible. For these applications, a support configuration as shown in Fig. 1.4 might yield better results [7]. Due to the shapes of the lowpass filters, this system is termed a diamond filter bank. In directional subband coding applications [3], [30], where directional sensitivity of the filters is important, the use of quadrant filter banks (Fig. 1.5) or filter banks with fan filters (Fig. 1.6) might be preferred. None of these support configurations, diamond, quadrant or fan, can be achieved by separable filter banks. Although nonseparable filter banks offer more flexibility and usually provide better performance, in most cases their design is considerably more difficult than separable filter banks. The implementation complexity of nonseparable filter banks is usually also higher.

Filter banks for the application of subband coding of speech were introduced in the 1970s [12]. Since then, studies on filter banks and subband coding have been booming [1]–[14],

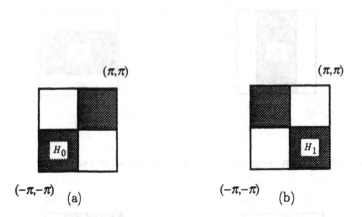

Figure 1.5. The support configuration of the quadrant filter bank. (a) The support of $H_0(\mathbf{w})$. (b) The support of $H_1(\mathbf{w})$.

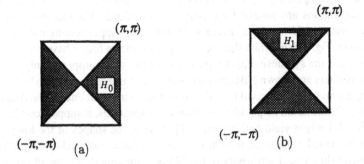

Figure 1.6. The filter bank with fan filters. (a) The support of $H_0(\mathbf{w})$. (b) The support of $H_1(\mathbf{w})$.

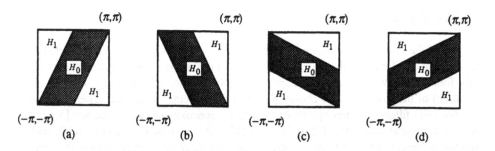

Figure 1.7. Four possible parallelogram supports of the two-channel filter bank.

[17]–[35], [37]–[39], [41]–[55]. Activities in the area of MD filter banks started in 1980s. Some of the earliest contributions were due to Vetterli [52], and Woods and O'Neil [54]. The idea of lattice decimation and expansion is an indispensable aspect of MD multirate systems. An introduction to MD sampling and signal processing can be found in [15]. A more detailed treatment is given in [50]. The theoretical aspects of MD systems are studied in [5]. An excellent tutorial of MD filter banks is given in [53]. Review of fundamentals of MD filter banks can be found at [48], [7], [Chapter 12, 50]. Results on the commutativity of MD decimators and expanders have been reported in [27], [22], [6], [17], [18]. Recently, the relation between filter banks and discrete wavelet transform has been extended to MD case [11], [28], and will not be elaborated in this paper.

1.1. Scope and Overview

The purpose of this paper is to provide an overview of developments on the design of MD filter banks, mostly 2D filter banks. Several design techniques for MD nonseparable filter banks will be discussed in this paper. They are outlined as follows.

1. Two-dimensional two-channel filter banks (Sec. II)

Commonly used 2D two-channel filter banks. Many studies have been done on 2D two-channel filter banks (Fig. 1.1 with $|\det \mathbf{M}| = 2$). In particular, the diamond filter bank (Fig. 1.4), first introduced by Vetterli [52], is of special interest in some image coding applications. The decimation matrix $\mathbf{M}$ for the diamond filter bank is usually the quincunx matrix $\mathbf{Q}$ as in (1.1). There is also some interest in the filter bank with quadrant filters as shown in Fig. 1.5, [3], [30], [45]. The support of $H_0(\mathbf{z})$ is in quadrants I and III while the support of $H_1(\mathbf{z})$ is in quadrants II and IV. Filter banks in which the filters have parallelogram supports are of importance in some applications [3]. Several possible parallelogram supports for the analysis and synthesis filters are shown in Fig. 1.7. The filter bank with fan filters (Fig. 1.6) is closely related to the diamond filter bank. From Fig. 1.6, we see that the fan filters can be obtained by shifting the diamond filters by $(\pi \quad 0)^T$ in frequency domain.

Design techniques

Design of the diamond filter bank. Most of the design techniques for 2D two-channel systems are developed for the diamond filter bank. For the two-channel systems, there are only four filters, two analysis filters and two synthesis filters. So in some designs, two (or three) of the four filters are chosen such that the system is alias free. The remaining two (or one) filters are then optimized to achieve approximate reconstruction [3], [52] or perfect reconstruction [1], [2], [24], [38], [45]. As 2D filters are considerably more difficult to design than 1D filters, in many cases (approximate reconstruction or perfect reconstruction) the 2D filters are obtained from 1D filters by appropriate mappings [1–3], [24], [38], [45]. In [2], Ansari and Lau proposed a design technique for the perfect reconstruction diamond filter bank. A polyphase mapping method is proposed therein to design IIR analysis filters. For filter banks with FIR filters, several 1D to 2D transformations have been considered. For example, the McClellan transformation is used in [1]. In [45], a more general transformation is considered and the design technique can be used for the diamond filter bank, the quadrant filter bank or filter banks with other types of supports. More recently [24], [38], a polyphase mapping similar to that in [2] is used to design a diamond filter bank. This technique allows the use of FIR or IIR filters, and moreover, in the IIR case the filters are guaranteed to be causal and stable.

Design of other types of filter banks. The design of the other commonly used filter banks is closely related to that of the diamond filter bank. The fan filters (Fig. 1.6) are shifted versions of the diamond filters and hence can be obtained by first designing the diamond filter bank. The filter banks in which the filters have parallelogram supports (Fig. 1.7) can be derived from the diamond filter bank by using the so-called unimodular transformation. This will be addressed in Sec. III. Some of the design techniques developed for the diamond filter bank [24], [38], [45], can be applied to the quadrant filter bank (Fig. 1.5) with some modifications.

In Sec. II, we will review two design techniques for the diamond filter bank. The first one proceeds along the line of [1] and the second one proceeds along the line of [24], [38]. A design example of the diamond filter bank through the second approach is also given in Sec. II. Although the design of quadrant filter banks (Fig. 1.5) is not mentioned in these references, we will see that the generalization to quadrant filter banks can be achieved easily.

2. Designs of MD multiple channel filter banks using transformations (Sec. III)

As MD nonseparable filter banks are considerably more difficult to design than 1D filter banks, various 1D to MD transformations have been proposed for designing suboptimal MD filter banks without actually optimizing MD filters. In the design of two-channel 2D filter banks (Sec. II), we will see such examples. For filter banks with more than two channels, two types of transformation have been proposed. In [49], the so-called unimodular transformation is developed. By use of the unimodular transformation, we can convert a MD separable filter bank to a nonseparable one. Consider a D-dimensional M-channel filter bank with separable filters. Applying unimodular transformation on the separable system, the new filter bank is still D-dimensional and has M channels, but the new analysis

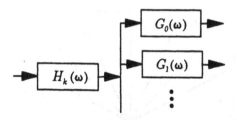

Figure 1.8. Tree structured analysis filter bank. Splitting of the *k*th subband.

and synthesis filters are nonseparable. Using the unimodular transformation, we will see that the 2D two-channel filter banks with parallelogram-supported filters (Fig. 1.7) can be derived from the diamond filter bank. In [43], Shah and Kalker studied transformations of 1D filter banks to D-dimensional filter banks. They proposed a 1D to D-dimensional transformation that preserves perfect reconstruction. We will present a review of these two transformations in Sec. III.

3. Tree-structured filter banks (Sec. IV)

Given an analysis filter bank, suppose we take a particular subband and split it into further subbands as shown in Fig. 1.8. By repeating this operation, we can actually build up a tree structured analysis bank. The most common example of a 1D tree structured filter bank is the one that results in a octave stacking of the passbands. In the 2D case, tree structures based on simple two-channel modules can offer sophisticated band-splitting schemes (sophisticated supports), especially if we combine the various configurations in Fig. 1.4–Fig. 1.7 appropriately. The directional filter bank developed by Bamberger and Smith (Fig. 1.9) is such an example [3]. We will review a number of tree structure examples in Sec. IV.

4. Two-dimensional cosine modulated filter bank (Sec. V–VII)

The one-dimensional (1D) cosine modulated filter bank (CMFB) has been studied extensively in the past [10], [41], [37], [39], [35], [26]. In the 1D CMFB, each analysis and synthesis filter is a cosine modulated version of one or two prototype filters. The CMFB has the advantages of low design cost and low implementation complexity. The success with 1D CMFB motivates the construction of 2D CMFB [19], [20], [29], [32], [31]. The separable 2D CMFB can always be obtained through concatenation of two 1D CMFB by using a tree structure. Our interest here is in designing a nonseparable 2D CMFB. The prototype filter is in general a nonseparable 2D FIR filter with a parallelogram support. Each analysis and synthesis filter is a cosine modulated version of the prototype, and is also nonseparable. In the separable 2D CMFB case, each individual filter consists of four shifts of a separable 2D prototype. However, the real-coefficient constraint on the analysis filter

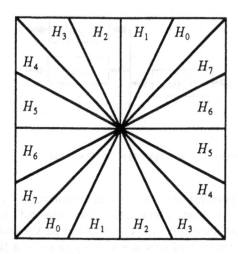

Figure 1.9. Supports of the analysis filters in the directional filter bank.

requires only two copies of the prototype. We can conceive that in the more general 2D CMFB the analysis filters can have two or four shifted copies of the prototype. So we will consider two classes of 2D FIR paraunitary cosine modulated filter banks: the two-copy CMFB and the four-copy CMFB. In the two-copy CMFB, each analysis filter contains two copies of the prototype and in the four-copy CMFB, each analysis filter contains four copies of the prototype. The filter bank will eventually be constrained to be paraunitary; the synthesis filters have the same support as the corresponding analysis filters.

The construction of the two-copy CMFB is analogous to the construction of 1D CMFB. For a filter bank with decimation matrix $\mathbf{M}$, non-diagonal in general, we start from a uniform 2D DFT filter bank with twice the number of channels. All the filters in the DFT filter bank is a shifted version of a prototype filter, which has a parallelogram support. The filters in the DFT filter bank are shifted and then paired to obtain real-coefficient analysis filters. Each analysis filter consists of two copies of the prototype and is a cosine modulated version of the prototype. We then study how to cancel major aliasing and constrain the prototype to ensure perfect reconstruction of the two-copy CMFB.

The four-copy CMFB will be constructed in a similar way. But in the four-copy case, we start from a uniform DFT filter bank with four times the number of channels. Then we shift the filters in the DFT filter bank and combine four shifted filters to obtain real-coefficient analysis filters. Necessary conditions on the decimation matrix will be derived for cancellation of major aliasing.

5. Two-dimensional FIR lossless systems (Sec. VIII)

In a M-channel paraunitary filter bank, the polyphase matrix of the analysis bank $\mathbf{E}(z)$ is paraunitary, [50]. If in addition to being paraunitary, $\mathbf{E}(z)$ is causal and stable, we say $\mathbf{E}(z)$ is lossless. Lossless matrices play an important role in the study and design of paraunitary systems. The 1D FIR lossless systems have been studied extensively in the past. It is known that 1D causal rational systems can be described by state space description. Moreover, the 1D FIR lossless systems can be characterized in terms of state space realization. The connection between the losslessness of the transfer matrix and the unitariness of the realization matrix is well-established [46], [50]. This connection also forms the basis for the factorization result developed in [14]. It is shown therein that the class of 1D FIR lossless systems can be factorized into some basic building blocks. To be more specific, we need to introduce the notion of degree for causal systems. Let $\mathbf{E}(z)$ be a causal system, then its degree is the minimum number of delay units (i.e., z^{-1}) required to implement $\mathbf{E}(z)$. It is shown in [14] that, all the 1D FIR lossless systems can be expressed as a product of a unitary matrix and degree-one building blocks. Similar approach can be attempted in the 2D case, but the details are more involved. An extensive study has been made by Venkataraman and Levy [51] and independently by Basu, Choi, and Chiang [4]. In Sec. VIII we will review some of the results in [51], [4].

1.2. Paper Outline and Notations

Paper outline. This paper is organized as follows: Sec. II is devoted to the designs of 2D two-channel filter banks, particularly the diamond filter bank and the quadrant filter bank. In Sec. III, we discuss the design of MD filter banks by using transformations. Some useful special cases of tree-structured filter banks are given in Sec. IV. In Sec. V, we introduce the two-copy CMFB and four-copy CMFB. Several issues that arise in the design of 2D CMFB will also be addressed. Details of two-copy CMFB and four-copy CMFB are given respectively in Sec. VI and VII. Developments in the study of 2D FIR lossless systems are presented in Sec. VIII. We will see the characterization of 2D FIR lossless systems in terms of state space realization. Factorizability of 2D FIR lossless systems will also be discussed in Sec. VIII. Except for the results presented in Sec. III, all the design techniques discussed in this paper are developed for 2D systems.

Notations. Notations in this paper are as in [50]. In particular, we will use the following notations.

(a) Boldfaced lower case letters are used to represent vectors and boldfaced upper case letters are reserved for matrices. The notations $\mathbf{A}^T$, $\mathbf{A}^*$, and $\mathbf{A}^\dagger$ represent the transpose, conjugate, and transpose-conjugate of $\mathbf{A}$. The 'tilde' notation is defined as follows: $\widetilde{\mathbf{A}}(z) = \mathbf{A}^\dagger(1/z^*)$.

(b) When a function has vector argument, e.g. $P(\mathbf{z})$, $\mathbf{z} = (\, z_0 \quad z_1 \quad \cdots \quad z_{D-1} \,)^T$, we will use $P(\mathbf{z})$ and $P(\, z_0 \quad z_1 \quad \cdots \quad z_{D-1} \,)$ interchangeably.

(c) The notation $\mathbf{I}_k$ denotes a $k \times k$ identity matrix, and the subscript will be omitted when it is clear from the context.

1.3. Summary of Integer Matrices and Multirate Systems

Fundamentals of integer matrices and multirate systems can be found in [7], [48], [50], [53]. A summary is given below.

1.3.1. Fundamentals of integer matrices

1. *Unimodular matrix.* An integer matrix U is unimodular if $|\det U| = 1$.

2. *The notations $\mathcal{N}(M)$, $J(M)$ and $SPD(M)$.* Let M be a $D \times D$ nonsingular integer matrix. The notation $\mathcal{N}(M)$ is defined as the set of integer vectors of the form Mx, where $x \in [0, 1)^D$. The number of elements in $\mathcal{N}(M)$ is denoted by $J(M)$, which is equal to $|\det M|$. In 1D case $D = 1$, and $\mathcal{N}(M) = \{0, 1, 2, \ldots, M - 1\}$. The symmetric parallelepiped $SPD(M)$ is defined as $SPD(M) = \{Mx, \ x \in [-1, 1)^D\}$.

3. *Division theorem for integer vectors.* Let M be a $D \times D$ matrix and n be a $D \times 1$ integer vector. We can express n as $n = n_0 + Mk$, $n_0 \in \mathcal{N}(M)$. Moreover, n_0 and k are unique. This relation is denoted by $n = n_0 \bmod M$.

4. *Lattices.* The lattice generated by an integer matrix M is denoted by $LAT(M)$.

5. *The Smith form.* A $D \times D$ integer matrix M can always be factorized as $M = U\Lambda V$, where U and V are unimodular integer matrices and Λ is a diagonal integer matrix. Furthermore, we can always ensure that the diagonal elements $[\Lambda]_{ii}$ of Λ are positive integers and $[\Lambda]_{ii}$ divides $[\Lambda]_{i+1,i+1}$. In this case, Λ is unique and is called the Smith form of M.

1.3.2. Basic notions in MD multirate systems

1. *Fourier transform and Z-transform.* Consider a D dimensional signal $x(n)$, where n is a $D \times 1$ integer column vector. The Fourier transformation and Z-transform of $x(n)$ will be denoted respectively by $X(w)$ and $X(z)$; they will be distinguished by the given argument, w for Fourier transform and z for Z-transform. The Fourier transform of $x(n)$ is defined as

$$X(w) = \sum_{n \in \mathcal{N}} x(n)e^{-jw^T n},$$

where w is a $D \times 1$ vector with $w = (\ \omega_0 \ \ \omega_1 \ \ \ldots \ \ \omega_{D-1}\ )^T$ and $\mathcal{N}$ is the set of all $D \times 1$ integer vectors. The Z-transform of $x(n)$, where it converges, is given by

$$X(z) = \sum_{n \in \mathcal{N}} x(n)z^{-n},$$

16

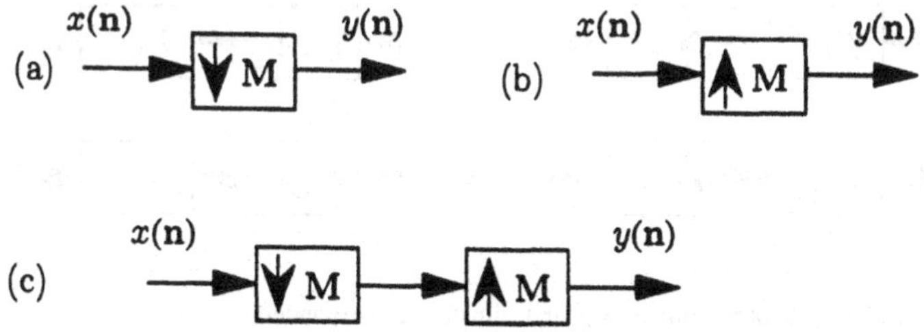

Figure 1.10. (a) The decimation matrix **M**, (b) the expansion matrix **M** and (c) the decimator followed by expander.

where $\mathbf{z} = (\ z_0\ \ z_1\ \ldots z_{D-1}\ )^T$. A vector raised to a vector power, as in $\mathbf{z}^{-\mathbf{n}}$ above, gives a scalar quantity defined as

$$\mathbf{z}^{\mathbf{n}} = z_0^{n_0} z_1^{n_1} \ldots z_{D-1}^{n_{D-1}}, \quad \mathbf{n} = (\ n_0\ \ n_1\ \ldots\ n_{D-1}\ )^T.$$

2. A filter $H(\mathbf{z})$ is called Nyquist($\mathbf{M}$) if $h(\mathbf{M}\mathbf{n})$ has only one nonzero coefficient. In 1D case, if $M = 2$, $H(z)$ is called a halfband filter.

3. *Decimators and expanders.* For an **M**-fold decimator (1.10(a)), the input $x(\mathbf{n})$ and the output $y(\mathbf{n})$ are related by $y(\mathbf{n}) = x(\mathbf{M}\mathbf{n})$. In the frequency domain, the relation is

$$Y(\mathbf{w}) = \frac{1}{J(\mathbf{M})} \sum_{\mathbf{k} \in \mathcal{N}(\mathbf{M}^T)} X(\mathbf{M}^{-T}(\mathbf{w} - 2\pi\mathbf{k})).$$

Given an **M**-fold expander (Fig. 1.10(b)), the input $x(\mathbf{n})$ and the output $y(\mathbf{n})$ are related by

$$y(\mathbf{n}) = \begin{cases} x(\mathbf{M}^{-1}\mathbf{n}), & \mathbf{n} \in LAT(\mathbf{M}) \\ 0, & \text{otherwise.} \end{cases}$$

In the frequency domain, the relation is $Y(\mathbf{w}) = X(\mathbf{M}^T\mathbf{w})$, or equivalently $Y(\mathbf{z}) = X(\mathbf{z}^{\mathbf{M}})$, where $\mathbf{z}^{\mathbf{M}}$ is defined as

$$\mathbf{z}^{\mathbf{M}} = (\ \mathbf{z}^{\mathbf{m}_0}\ \ \mathbf{z}^{\mathbf{m}_1}\ \ldots\ \mathbf{z}^{\mathbf{m}_{D-1}}\ )^T,$$

with $\mathbf{m}_i$ denoting the ith column of **M**.

4. *Decimation followed by expansion.* Consider the concatenation of the **M**-fold decimator and the **M**-fold expander in Fig. 1.10(c). The input $x(\mathbf{n})$ and the output $y(\mathbf{n})$ are related by

$$y(\mathbf{n}) = \begin{cases} x(\mathbf{n}), & \mathbf{n} \in LAT(\mathbf{M}) \\ 0, & \text{otherwise.} \end{cases}$$

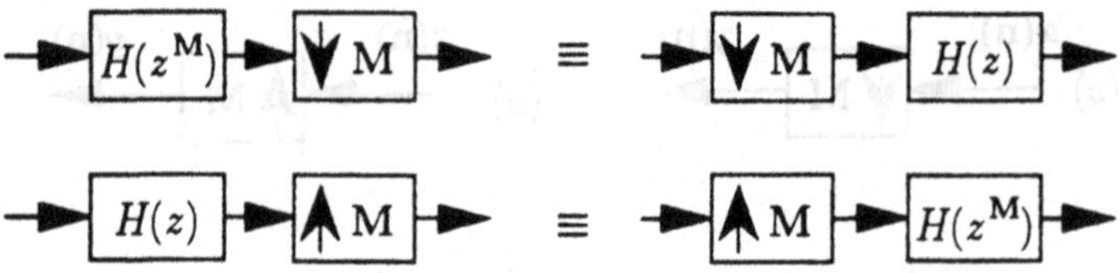

Figure 1.11. Noble identities for multidimensional decimators and expanders.

In the frequency domain, the relation becomes

$$Y(\mathbf{w}) = \frac{1}{J(\mathbf{M})} \sum_{\mathbf{k} \in \mathcal{N}(\mathbf{M}^T)} X(\mathbf{w} - 2\pi \mathbf{M}^{-T} \mathbf{k}).$$

The output $Y(\mathbf{w})$ contains $X(\mathbf{w})$ and $J(\mathbf{M}) - 1$ shifted versions $X(\mathbf{w} - 2\pi \mathbf{M}^{-T} \mathbf{k})$ (images of $X(\mathbf{w})$).

5. *Noble identities.* Fig. 1.11 shows two useful multirate identities for multidimensional systems. These allow the movement of multirate building blocks across transfer functions under some conditions.

6. *Perfect reconstruction MD filter bank.* Consider the MD filter bank in Fig. 1.1. Let $\mathcal{N}(\mathbf{M}^T) = \{ \mathbf{k}_i \}_{i=0}^{J(\mathbf{M})-1}$ and the vector $\mathbf{k}_0 = \underline{0}$. The output $\widehat{X}(\mathbf{w})$ is given by

$$\widehat{X}(\mathbf{w}) = T(\mathbf{w})X(\mathbf{w}) + \sum_{i=1}^{J(\mathbf{M})-1} A_i(\mathbf{w})X(\mathbf{w} - 2\pi \mathbf{M}^{-T} \mathbf{k}_i), \qquad (1.2)$$

where $T(\mathbf{w})$ is the distortion function and $A_i(\mathbf{w})$ is the ith aliasing transfer function. The distortion function $T(\mathbf{w})$ is defined as

$$T(\mathbf{w}) = \frac{1}{J(\mathbf{M})} \sum_{m=0}^{J(\mathbf{M})-1} H_m(\mathbf{w}) F_m(\mathbf{w}).$$

The ith aliasing transfer function $A_i(\mathbf{w})$ is defined as

$$A_i(\mathbf{w}) = \frac{1}{J(\mathbf{M})} \sum_{m=0}^{J(\mathbf{M})-1} H_m(\mathbf{w} - 2\pi \mathbf{M}^{-T} \mathbf{k}_i) F_m(\mathbf{w}).$$

The MD filter bank is free from aliasing if $A_i(\mathbf{w}) = 0$, for $i = 1, 2, \cdots, J(\mathbf{M}) - 1$. The filter bank has perfect reconstruction if it is free from aliasing and the distortion function $T(\mathbf{w})$ is a delay. In this case, $\widehat{X}(\mathbf{w})$ is a scaled and delayed version of $X(\mathbf{w})$. As in 1D filter banks, the perfect reconstruction condition can be interpreted in terms

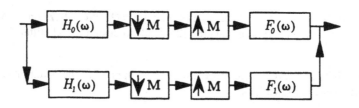

Figure 2.1. Two-dimensional two-channel filter bank, where **M** is a 2 by 2 integer matrix with | det **M**|=2.

of the polyphase matrices. Using polyphase decomposition, the analysis and synthesis filters have the form

$$H_m(\mathbf{w}) = \sum_{\mathbf{n}_i \in \mathcal{N}(\mathbf{M})} E_{m,i}(\mathbf{M}^T \mathbf{w}) e^{-j\mathbf{w}^T \mathbf{n}_i}, \quad m = 0, 1, \ldots, J(\mathbf{M}) - 1,$$

$$F_m(\mathbf{w}) = \sum_{\mathbf{n}_i \in \mathcal{N}(\mathbf{M})} R_{i,m}(\mathbf{M}^T \mathbf{w}) e^{j\mathbf{w}^T \mathbf{n}_i}, \quad m = 0, 1, \ldots, J(\mathbf{M}) - 1.$$

The $J(\mathbf{M}) \times J(\mathbf{M})$ matrices $\mathbf{E}(\mathbf{w})$ and $\mathbf{R}(\mathbf{w})$ with $[\mathbf{E}(\mathbf{w})]_{m,i} = E_{m,i}(\mathbf{w})$ and $[\mathbf{R}(\mathbf{w})]_{m,i} = R_{m,i}(\mathbf{w})$ are respectively called the polyphase matrices for the analysis bank and the synthesis bank. The MD filter bank has perfect reconstruction if $\mathbf{R}(\mathbf{w})\mathbf{E}(\mathbf{w}) = \mathbf{I}$.

2. Two-Dimensional Two-Channel Filter Banks

Consider the 2D two-channel filter bank in Fig. 2.1, where the decimation matrix **M** is a 2×2 integer matrix with $|\det \mathbf{M}| = 2$. For the diamond filter bank, the decimation matrix is usually the quincunx matrix **Q** as defined in (1.1). The supports of the analysis filters are shown in Fig. 1.4. As there are only four filters in the filter bank, two analysis and two synthesis filters, it is possible to design perfect reconstruction diamond filter bank without direct optimization of 2D filters. The design can be first reduced to that of two 2D filters, which can be further reduced to the design of 1D filters. More precisely, we can design the diamond filter bank by the following two steps.

1. *Reducing the design of the diamond filter bank to that of two 2D filters* [1]. If we choose $H_1(\mathbf{z})$ and $F_1(\mathbf{z})$ appropriately, we can reduce the design of the diamond filter bank to that of $H_0(\mathbf{z})$ and $F_0(\mathbf{z})$. In this case, the filter bank has perfect reconstruction if and only if $H_0(\mathbf{z}) F_0(\mathbf{z})$ is Nyquist(**Q**). (The definition of a Nyquist(**Q**) filter is as given in Sec. 1.3.2.) So the design of the diamond filter bank is reduced to the design of 2D filters $H_0(\mathbf{z})$ and $F_0(\mathbf{z})$.

2. *Design of $H_0(\mathbf{z})$ and $F_0(\mathbf{z})$ such that $H_0(\mathbf{z}) F_0(\mathbf{z})$ is Nyquist(**M**).* Two approaches can be used to carry out this step. Both design techniques involve only the design of 1D filters; no 2D optimization is required. In the first approach [1], the McClellan transformation is employed to convert the 2D filter design problem to a similar 1D problem. In the second approach [24], [38], we will use a Nyquist(**M**) approach to further simplify the

19

design of $H_0(z)$ and $F_0(z)$ to the design of a Nyquist($\mathbf{Q}$) filter $H_0(z)$ only. After this, we use the so-called polyphase mapping [24], [38] to design the 2D filter $H_0(z)$ from a 1D filter.

We will see that the above design procedures can be applied to the quadrant filter bank (Fig. 1.5) with some modifications.

2.1. Design of the Diamond Filter Bank

2.1.1. Reducing the filter bank design problem to a constrained 2D filter-design problem

Consider the filter bank in Fig. 2.1 with decimation matrix $\mathbf{M} = \mathbf{Q}$, where $\mathbf{Q}$ is as defined in (1.1). With this choice of $\mathbf{M}$, it can be verified from (1.2) that the output $\widehat{X}(z)$ and input $X(z)$ are related by $\widehat{X}(z) = A(z)X(-z) + T(z)X(z)$, where $A(z)$ and $T(z)$ are respectively the alias transfer function and the distortion function given by

$$A(z) = \frac{1}{2}(H_0(-z)F_0(z) + H_1(-z)F_1(z))$$

$$T(z) = \frac{1}{2}(H_0(z)F_0(z) + H_1(z)F_1(z)). \tag{2.1}$$

The system is alias free if the alias transfer function $A(z) = 0$. The system has perfect reconstruction if it is alias free and the distortion function $T(z)$ is a delay. For alias cancellation, we choose the following highpass analysis and synthesis filters

$$H_1(z) = z_0^{-1}F_0(-z), \qquad F_1(z) = z_0 H_0(-z). \tag{2.2}$$

With this choice, the alias cancellation condition $A(z) = 0$ is satisfied. Furthermore, $H_1(z)$ and $F_1(z)$ will have the desired support if $H_0(z)$ and $F_0(z)$ have the desired support of the lowpass filters in Fig. 1.4. In this case the distortion function is

$$T(z) = \frac{1}{2}(H_0(z)F_0(z) + H_0(-z)F_0(-z)),$$

which is a delay if and only if $H_0(z)F_0(z)$ is a Nyquist($\mathbf{Q}$) filter.

So the design of perfect reconstruction diamond filter bank reduces to the design of diamond-shaped $H_0(z)$ and $F_0(z)$ such that $H_0(z)F_0(z)$ is a Nyquist($\mathbf{Q}$) filter.

2.1.2. Design of $H_0(z)$ and $F_0(z)$ such that $H_0(z)F_0(z)$ is Nyquist($\mathbf{Q}$): method 1

In method 1, we will use the McClellan transformation to simplify the design problem of 2D filters $H_0(z)$ and $F_0(z)$. In particular, we can convert the design of $H_0(z)$ and $F_0(z)$ into a similar 1D problem: design of 1D filters $H(z)$ and $F(z)$ such that $H(z)F(z)$ is halfband (i.e., Nyquist(2)). The analysis and synthesis filters in this case are FIR and linear-phase.

McClellan transformation. This transformation was proposed by McClellan in 1976 and has been since a popular method in designing 2D filters [15], [36]. It converts a

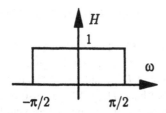

Figure 2.2. The desired response of $H(\omega)$.

1D zero phase filter into a 2D zero phase filter. The procedure of applying McClellan transformation on a 1D zero phase filter $P(z)$ is as follows. Any 1D zero phase filter $P(z)$ with real coefficients can be expressed in the form $P(z) = \sum_{n=0}^{K} \hat{p}(n)(z + z^{-1})^n$, where $\hat{p}(n)$ is real. Thus $P(z) = \widehat{P}(z + z^{-1})$, where $\widehat{P}(x)$ is the polynomial $\widehat{P}(x) = \sum_{n=0}^{K} \hat{p}(n)x^n$. Given such a 1D filter $P(z)$, consider the 2D filter defined from the underlying polynomial $\widehat{P}(x)$ as follows

$$P_0(z_0, z_1) = \widehat{P}(M(z_0, z_1)), \quad \text{where}$$
$$M(z_0, z_1) = a_0 + a_1(z_0 + z_0^{-1}) + a_2(z_1 + z_1^{-1})$$
$$+ a_3(z_0 z_1^{-1} + z_0^{-1} z_1) + a_4(z_0 z_1 + z_0^{-1} z_1^{-1}). \tag{2.3}$$

With different choice of a_i, the 2D filter $P_0(\mathbf{z})$ has different frequency response.

For our purpose, the appropriate choice of $M(z_0, z_1)$ is $M_d(z_0, z_1)$, where

$$M_d(z_0, z_1) = \frac{1}{2}(z_0 + z_0^{-1} + z_1 + z_1^{-1}). \tag{2.4}$$

If the 1D filter $P(z)$ is lowpass as shown in Fig. 2.2, then $P_0(\mathbf{z})$ has a diamond-shaped support as shown in Fig. 1.4(a). Furthermore, it can be verified that if the 1D filter $P(z)$ is half-band, then the 2D filter $P_0(\mathbf{z})$ is Nyquist($\mathbf{Q}$).

The two steps involved in designing the analysis bank $\{H_0(\mathbf{z}), F_0(\mathbf{z})\}$ are:

1. Design 1D zero-phase filters $H(z)$ and $F(z)$ such that $D(z) = H(z)F(z)$ is halfband.

2. Apply the McClellan transformation $M_d(\mathbf{z})$ on $H(z)$ and $F(z)$ to get the 2D filters $H_0(\mathbf{z})$ and $F_0(\mathbf{z})$.

If the 1D filters $H(z)$ and $F(z)$ are lowpass as in Fig. 2.2, the 2D filters $H_0(\mathbf{z})$ and $F_0(\mathbf{z})$ have the desired diamond support shown in Fig. 1.4(a). As $H(z)F(z)$ is halfband, the product filter $H_0(\mathbf{z})F_0(\mathbf{z})$ is Nyquist($\mathbf{Q}$); perfect reconstruction is guaranteed. Because the 2D filters designed through the McClellan transformation are FIR and linear-phase, all the analysis and synthesis filters are FIR and linear-phase. We observe that the remaining nontrivial part of this technique is the design of 1D zero-phase filters $H(z)$ and $F(z)$ such that $H(z)F(z)$ is halfband. The design of such $H(z)$ and $F(z)$ is discussed next.

Design of 1D zero-phase filters $H_0(\mathbf{z})$ and $F_0(\mathbf{z})$ such that $H_0(\mathbf{z})F_0(\mathbf{z})$ is halfband

We will discuss two approaches for this. These two approaches are elaborated below.

1. Factorization approach [1]. We can first design a zero-phase halfband $D(z)$ and then factorize $D(z)$ into linear phase $H(z)$ and $F(z)$. The technique described in [47] can be used to design linear-phase halfband $D(z)$. We can also choose $D(z)$ to be a Lagrange filter [45], which is a well-known class of zero-phase halfband filters that can be expressed in closed form. However, this design of $H(z)$ and $F(z)$ involves factorization of $D(z)$. We will see that the Nyquist(2) approach to be introduced next requires no factorization.

2. Nyquist(2) approach [25], [38]. Suppose $H(z)$ itself is a Nyquist(2) (i.e., halfband) filter. It can be verified if $H(z)F(z)$ has to be halfband, then $F(z)$ is necessarily of the form $F(z) = 1 + (2\gamma(z) - 1)H(-z)$, for some halfband filter $\gamma(z)$. Conversely, $H(z)F(z)$ is a halfband filter for any choice of halfband $H(z)$ and $\gamma(z)$. For simplicity, let us choose $\gamma(z) = H(z)$. Then

$$F(z) = 1 + (2H(z) - 1)H(-z).$$

It follows that if $H(z)$ is lowpass, $F(z)$ is also lowpass. Zero-phase property of $H(z)$ implies zero-phase property of $F(z)$. Through this approach, we only need to design $H(z)$, which can be easily done using any design technique for linear phase halfband filters. This Nyquist(2) approach can be extended to 2D case and the 2D extension will be the first step for the second design method of $H_0(\mathbf{z})$ and $F_0(\mathbf{z})$.

2.1.3. *Design of $H_0(\mathbf{z})$ and $F_0(\mathbf{z})$ such that $H_0(\mathbf{z})F_0(\mathbf{z})$ is Nyquist(Q): method 2*

This method can be described by the following two steps, [24], [38].
Step 1. Use a Nyquist(Q) approach to simplify the design of $H_0(\mathbf{z})$ and $F_0(\mathbf{z})$ to only the design of a Nyquist(Q) filter $H_0(\mathbf{z})$. Step 2. Design a Nyquist(Q) filter $H_0(\mathbf{z})$ with a diamond support.

The resulting filter bank has some very attractive properties. The individual filters can be FIR or IIR. Detailed discussion will be given in Sec. 2.3.

Step 1. Nyquist(Q) approach. Let $H_0(\mathbf{z})$ be a Nyquist(Q) filter and $H_0(\mathbf{z}) + H_0(-\mathbf{z}) = 1$. We can verify that if $H_0(\mathbf{z})F_0(\mathbf{z})$ is also Nyquist(Q), then $F_0(\mathbf{z})$ is necessarily of the form $F_0(\mathbf{z}) = 1 + (2\gamma(\mathbf{z}) - 1)H_0(-\mathbf{z})$, for some Nyquist(Q) filter $\gamma(\mathbf{z})$. For simplicity, we can choose $\gamma(\mathbf{z}) = H_0(\mathbf{z})$, then

$$F_0(\mathbf{z}) = 1 + (2H_0(\mathbf{z}) - 1)H_0(-\mathbf{z}).$$

For this choice of $F_0(\mathbf{z})$, the product filter $H_0(\mathbf{z})F_0(\mathbf{z})$ is Nyquist(Q) if $H_0(\mathbf{z})$ is Nyquist(Q). So perfect reconstruction of the diamond filter bank is guaranteed if $H_0(\mathbf{z})$ is Nyquist(Q). As Nyquist(Q) is the only condition on $H_0(\mathbf{z})$, $H_0(\mathbf{z})$ can be FIR or IIR. Furthermore, if $H_0(\mathbf{z})$ has a diamond-shaped support as shown in Fig. 1.4(a), $F_0(\mathbf{z})$ has the same diamond-shaped support. So the remaining task is to design $H_0(\mathbf{z})$ such that it is Nyquist(Q) and has the diamond support in Fig. 1.4(a).

Step 2. Design of Nyquist(Q) $H_0(z)$ with a diamond support. The McClellan transformation can be used to design such $H_0(z)$. Let $H(z)$ be a 1D zero-phase halfband filter with support as shown in Fig. 2.2. By using the McClellan transformation in (2.4) on $H(z)$, we can obtain a Nyquist(Q) filter $H_0(z)$ with a diamond support. Because the 2D filters designed through McClellan transformation are FIR and have linear phase, all the analysis and synthesis filters are FIR and are constrained to have linear phase. Next, we use a polyphase mapping method to design $H_0(z)$. In this method, $H_0(z)$ can be FIR or IIR and need not have linear phase in FIR case.

Polyphase mapping approach. As $H_0(z)$ is Nyquist(Q), without loss of generality, it can be expressed as

$$H_0(z) = \frac{1}{2}(1 + z_0^{-1}\beta(z^Q)).$$

In this case, $H_0(z)$ remains Nyquist(Q) for any choice of $\beta(z)$. Also let $H(z)$ be a 1D halfband filter with support as shown in Fig. 2.2, then $H(z)$ can be expressed as

$$H(z) = \frac{1}{2}(1 + z^{-1}\alpha(z^2)). \tag{2.5}$$

We will see that if we choose

$$\beta(z) = \alpha(z_0)\alpha(z_1), \tag{2.6}$$

then $H_0(z)$ has the desired diamond support. The 1D polyphase component $\alpha(z)$ can be FIR or IIR and hence the 2D filter $H_0(z)$ can be FIR or IIR. As $\alpha(z)$ is the polyphase component of the 1D filter $H(z)$ and $\beta(z)$ is the polyphase component of the 2D filter $H_0(z)$, this method is termed the polyphase mapping method. This mapping can also be obtained by starting from the transformation reported in [9].

The reason why the mapping in (2.6) gives the desired response for $H_0(z)$ is as follows. Notice that if $H(z)$ is lowpass with support as shown in Fig. 2.2, then

$$z^{-1}\alpha(z^2) \approx \begin{cases} 1, & \omega \in (-\pi/2, \pi/2) \\ -1, & \text{otherwise.} \end{cases} \quad \text{(see Fig. 2.3)} \tag{2.7}$$

It follows that

$$(z_0 z_1)^{-1/2}\alpha(z_0 z_1) \approx \begin{cases} 1, & (\omega_0 + \omega_1) \in (-\pi, \pi), \\ -1, & \text{otherwise,} \end{cases} \quad \text{(see Fig. 2.4(a))}$$

$$\text{and} \quad (z_0 z_1^{-1})^{-1/2}\alpha(z_0 z_1^{-1}) \approx \begin{cases} 1, & (\omega_0 - \omega_1) \in (-\pi, \pi), \\ -1, & \text{otherwise.} \end{cases} \quad \text{(see Fig. 2.4(b))}$$

So $z_0^{-1}\beta(z^Q) = z_0^{-1}\alpha(z_0 z_1)\alpha(z_0 z_1^{-1})$, which is the product of $(z_0 z_1)^{-1/2}\alpha(z_0 z_1)$ and $(z_0 z_1^{-1})^{-1/2}\alpha(z_0 z_1^{-1})$, has the following response

$$z_0^{-1}\beta_0(z^Q) \approx \begin{cases} 1, & \mathbf{w} \in \text{the diamond region in Fig. 1.4(a).} \\ -1, & \text{otherwise.} \end{cases} \quad \text{(see Fig. 2.4(c))}$$

Fig. 2.4(c) implies that $H_0(\mathbf{w})$ has the desired diamond support.

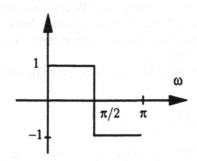

Figure 2.3. The ideal response of $z^{-1}\alpha(z^2)$.

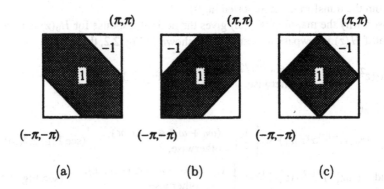

Figure 2.4. The ideal responses of (a) $(z_0z_1)^{-1/2}\alpha(z_0z_1)$, (b) $(z_0z_1^{-1})^{-1/2}\alpha(z_0z_1^{-1})$, and (c) $z_0^{-1}\beta(\mathbf{z}^{\mathbf{Q}})$ for the diamond filter bank.

2.2. Design of the Quadrant Filter Bank

2.2.1. Reducing the filter bank design to a constrained 2D filter-design problem

For the quadrant filter bank, the decimation matrix will be denoted as $\mathbf{D}$, where

$$\mathbf{D} = \begin{pmatrix} 2 & 0 \\ 0 & 1 \end{pmatrix}. \tag{2.8}$$

Consider the filter bank in Fig. 2.1 with decimation matrix $\mathbf{M} = \mathbf{D}$. The output $\widehat{X}(\mathbf{z})$ and input $X(\mathbf{z})$ in this case are related by $\widehat{X}(\mathbf{z}) = A(\mathbf{z})X(-z_0, z_1) + T(\mathbf{z})X(\mathbf{z})$. The alias transfer function $A(\mathbf{z})$ is given by

$$A(\mathbf{z}) = \frac{1}{2} \left(H_0(-z_0, z_1) F_0(\mathbf{z}) + H_1(-z_0, z_1) F_1(\mathbf{z}) \right).$$

The distortion function $T(\mathbf{z})$ is as given in (2.1). Consider the following choice of the filters $H_1(\mathbf{z})$ and $F_1(\mathbf{z})$.

$$H_1(\mathbf{z}) = z_0^{-1} F_0(-z_0, z_1), \qquad F_1(\mathbf{z}) = z_0 H_0(-z_0, z_1).$$

This choice gives us exact cancellation of aliasing, i.e. $A(\mathbf{z}) = 0$. Also $H_1(\mathbf{z})$ and $F_1(\mathbf{z})$ have the desired support as in Fig. 1.5(b) if $H_0(\mathbf{z})$ and $F_0(\mathbf{z})$ have supports in quadrant I and III as in Fig. 1.5(a). For the above choice of filters, the distortion function is given by

$$T(\mathbf{z}) = \frac{1}{2} \left(H_0(\mathbf{z}) F_0(\mathbf{z}) + H_0(-z_0, z_1) F_0(-z_0, z_1) \right).$$

The quadrant filter bank has perfect reconstruction if and only if $H_0(\mathbf{z}) F_0(\mathbf{z})$ is Nyquist($\mathbf{D}$). Similar to the diamond case, the design problem of the filter bank reduces to the design of quadrant filters $H_0(\mathbf{z})$ and $F_0(\mathbf{z})$ such that $H_0(\mathbf{z}) F_0(\mathbf{z})$ is a Nyquist($\mathbf{D}$) filter.

2.2.2. Design of $H_0(z)$ and $F_0(z)$ such that $H_0(z) F_0(z)$ is Nyquist($\mathbf{D}$): method 1

In the diamond filter bank, we have used a special case of McClellan transformation, namely (2.4), to design a 2D diamond-shaped filter. Here we will consider a different choice of McClellan transformation that converts a 1D zero-phase lowpass filter into a 2D quadrant filter. Let $P(z)$ be a 1D zero-phase filter with support as in Fig. 2.2. Consider the McClellan transformation

$$M_q(\mathbf{z}) = \frac{1}{2} \left(z_0 z_1 + z_0^{-1} z_1^{-1} - z_0 z_1^{-1} - z_0^{-1} z_1 \right). \tag{2.9}$$

If we apply this transformation on $P(z)$, the resulting 2D filter $P_0(\mathbf{z})$ has quadrant support as shown in Fig. 1.5(a). Moreover, if the 1D filter $P(z)$ is halfband, then $P_0(\mathbf{z})$ is Nyquist($\mathbf{D}$). As in the diamond case, let $H(z)$ and $F(z)$ have support as in Fig. 2.2 and let $H(z)F(z)$ be a halfband filter. Then we apply the McClellan transformation described in (2.9) on

$H(z)$ and $F(z)$ to obtain 2D filters $H_0(\mathbf{z})$ and $F_0(\mathbf{z})$. The resulting $H_0(\mathbf{z})$ and $F_0(\mathbf{z})$ are quadrant filters as in Fig. 1.5(a). As $H(z)F(z)$ is halfband, $H_0(\mathbf{z})F_0(\mathbf{z})$ is a Nyquist($\mathbf{D}$) filter; perfect reconstruction of the quadrant filter bank is assured. Also, as the 2D filters designed through the McClellan transformation are FIR and linear-phase, all the analysis and synthesis filters are FIR and linear-phase.

2.2.3. Design of $H_0(z)$ and $F_0(z)$ such that $H_0(z)F_0(z)$ is Nyquist($\mathbf{D}$): method 2

This technique is very similar to the one proposed for the diamond filter bank in Sec. 2.1.3. In this design, the individual filters can be FIR or IIR. This method can be described by the following two steps. Step 1. Use a Nyquist($\mathbf{D}$) approach to simplify the design of $H_0(\mathbf{z})$ and $F_0(\mathbf{z})$ to only the design of a Nyquist($\mathbf{D}$) filter $H_0(\mathbf{z})$. Step 2. Design a Nyquist($\mathbf{D}$) filter $H_0(\mathbf{z})$ with the quadrant support in Fig. 1.5(a).

Step 1. Nyquist($\mathbf{D}$) approach. Let $H_0(\mathbf{z})$ be a Nyquist($\mathbf{D}$) filter and $H_0(\mathbf{z}) + H_0(-z_0, z_1) = 1$. We can verify that if $H_0(\mathbf{z})F_0(\mathbf{z})$ is Nyquist($\mathbf{D}$), then $F_0(\mathbf{z})$ is of the form $F_0(\mathbf{z}) = 1 + (2\gamma(\mathbf{z}) - 1)H_0(-z_0, z_1)$, for some Nyquist($\mathbf{D}$) filter $\gamma(\mathbf{z})$. For simplicity, let us choose $\gamma(\mathbf{z}) = H_0(\mathbf{z})$, then

$$F_0(\mathbf{z}) = 1 + (2H_0(\mathbf{z}) - 1)H_0(-z_0, z_1).$$

For this choice of $F_0(\mathbf{z})$, the product filter $H_0(\mathbf{z})F_0(\mathbf{z})$ remains Nyquist($\mathbf{D}$) for any Nyquist($\mathbf{D}$) filter $H_0(\mathbf{z})$. So perfect reconstruction of the quadrant filter bank is guaranteed if $H_0(\mathbf{z})$ is Nyquist($\mathbf{Q}$). The only condition on $H_0(\mathbf{z})$ is that $H_0(\mathbf{z})$ should be Nyquist($\mathbf{D}$); $H_0(\mathbf{z})$ can be FIR or IIR. Furthermore, if $H_0(\mathbf{z})$ has a quadrant support as shown in Fig. 1.5(a), then $F_0(\mathbf{z})$ has the same quadrant support. The remaining task is to design the Nyquist($\mathbf{D}$) filter $H_0(\mathbf{z})$ with support as in Fig. 1.5(a).

Step 2. Design of Nyquist($\mathbf{D}$) $H_0(\mathbf{z})$. Similar to the design of filter bank with diamond filters, the McClellan transformation can be used to design $H_0(\mathbf{z})$. Let the 1D filter $H(z)$ be linear-phase and has support as in Fig. 2.2. By using the McClellan transformation in (2.9) on $H(z)$, we can obtain a Nyquist($\mathbf{D}$) filter $H_0(\mathbf{z})$ with a quadrant support as in (Fig. 1.5(a)). In this case, all the analysis and synthesis filters are FIR and are constrained to have linear phase. Next we design $H_0(\mathbf{z})$ using a polyphase mapping method, which is similar to the one introduced for the diamond filter in (2.6). In this method, $H_0(\mathbf{z})$ can be FIR or IIR and need not have linear phase in FIR case.

Polyphase mapping. As $H_0(\mathbf{z})$ is Nyquist($\mathbf{D}$), without loss of generality, $H_0(\mathbf{z})$ assumes the form

$$H_0(\mathbf{z}) = \frac{1}{2}(1 + z_0^{-1}\beta(\mathbf{z}^{\mathbf{D}})).$$

Let $H(z)$ be a 1D halfband filter as (2.5) and the support of $H(z)$ be as shown in Fig. 2.2. We will see that if we choose

$$\beta(\mathbf{z}) = -z_1^{-1}\alpha(-z_0)\alpha(-z_1^2), \tag{2.10}$$

then $H_0(\mathbf{z})$ has the desired quadrant support. This polyphase mapping is similar to the one proposed in (2.6) for diamond filters. In this case, the 1D polyphase component $\alpha(z)$ can

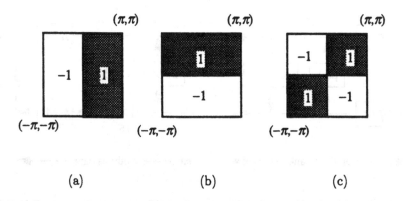

Figure 2.5. The ideal responses of (a) $jz_0\alpha(-z_0^2)$, (b) $jz_1\alpha(-z_1^2)$, and (c) $z_0^{-1}\beta(\mathbf{z^D})$ for the quadrant filter bank.

also be FIR or IIR and hence the 2D filter $H_0(\mathbf{z})$ can be FIR or IIR. The reason why the mapping in (2.10) gives the desired response is given next.

From (2.7), we have

$$(jz_0)\alpha(-z_0^2) \approx \begin{cases} 1, & \omega_0 \in (0, \pi), \\ -1, & \text{otherwise}, \end{cases} \quad \text{(see Fig. 2.5(a))}$$

$$(jz_1)\alpha(-z_1^2) \approx \begin{cases} 1, & \omega_1 \in (0, \pi), \\ -1, & \text{otherwise}, \end{cases} \quad \text{(see Fig. 2.5(b))}$$

Therefore, $z_0^{-1}\beta(\mathbf{z^D}) = -z_0^{-1}z_1^{-1}\alpha(-z_0^2)\alpha(-z_1^2)$ has the following response

$$z_0^{-1}\beta_0(\mathbf{z^D}) \approx \begin{cases} 1, & \mathbf{w} \in \text{quadrants I and III}, \\ -1, & \text{otherwise}. \end{cases} \quad \text{(see Fig. 2.5(c))}$$

Fig. 2.5(c) implies that $H_0(\mathbf{z})$ has the desired support.

2.3. *Properties due to the second design technique of $H_0(\mathbf{z})$ and $F_0(\mathbf{z})$*

For both the diamond and the quadrant filter banks, we have described two techniques to design $H_0(\mathbf{z})$ and $F_0(\mathbf{z})$ such that $H_0(\mathbf{z})F_0(\mathbf{z})$ is Nyquist($\mathbf{M}$). When the second design technique (Sec. 2.1.3 for the diamond filter bank and Sec. 2.2.3 for the quadrant filter bank) is used, the filter bank has some very attractive properties.

Robust ladder structure. When Nyquist($\mathbf{M}$) approach is used to design $H_0(\mathbf{z})$ and $F_0(\mathbf{z})$, the proposed filter bank has a very attractive ladder structure implementation (Fig. 2.6), which is robust to roundoff noise [38]. For the diamond filter bank, $\beta(\mathbf{z})$ is as in (2.6), and $\mathbf{d} = (\ 1 \ \ 1 \ )^T$. For the quadrant filter bank, $\beta(\mathbf{z})$ is as in (2.10), and $\mathbf{d} = (\ 1 \ \ 0 \ )^T$. Perfect reconstruction is preserved even when coefficients of $\beta(\mathbf{z})$ are quantized.

If we further use the polyphase mapping method to design the Nyquist($\mathbf{M}$) filter $H_0(\mathbf{z})$, the filter bank has the following additional three properties.

27

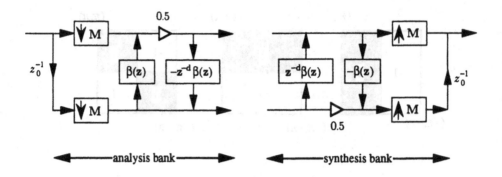

Figure 2.6. The implementation of the diamond filter bank or the quadrant filter bank that is designed through the Nyquist(M) approach.

1. *Stability and causality in IIR case.* Because the the 1D filter $H(z)$ in (2.5) can be FIR or IIR, its polyphase component $\alpha(z)$ can be FIR or IIR. So the 2D analysis filters can be FIR or IIR. In the IIR case, it is shown in [38] that the 2D analysis and synthesis filters will always be causal and stable if $\alpha(z)$ is causal and stable. Also in the IIR case, $\alpha(z)$ can be taken to be an allpass function. The allpass functions can be implemented through a low-sensitivity lattice structure [50], which guarantees stability in spite of multiplier quantization.

2. *Linear-phase in FIR case.* If the 1D filter $\alpha(z)$ is a Type 2 linear-phase filter [50], the analysis and synthesis filters have linear phase.

3. *Complexity.* From the implementation in Fig. 2.6, we observe that the complexity of the analysis bank is comparable to that of $\beta(\mathbf{z})$ (due to the fact that all the operations are at a lower rate). For both diamond and quadrant filter banks, $\beta(\mathbf{z})$ is a separable filter and the cost of $\beta(\mathbf{z})$ is equivalent to twice that of the 1D filter $\alpha(z)$.

Example 2.1. *FIR diamond filter bank.* In this example, we use the technique described in Sec. 2.1.3 to design $H_0(\mathbf{z})$ and $F_0(\mathbf{z})$. We first design 1D FIR halfband filter $H(z)$ that has zero phase and has support as in Fig. 2.2. Then we use the polyphase mapping method to obtain the 2D Nyquist(Q) filter $H_0(\mathbf{z})$ from $H(z)$. As $H_0(\mathbf{z})$ is Nyquist(Q), the diamond filter bank has perfect reconstruction. Due to the polyphase mapping method, $H_0(\mathbf{z})$ is FIR and linear-phase and hence all the analysis and synthesis filters are FIR and linear-phase. Fig. 2.7(a) and Fig. 2.7(b) show respectively the magnitude responses of $H_0(\mathbf{z})$ and $H_1(\mathbf{z})$. The stopband attenuation $\delta(H_0) \approx 40$ dB and $\delta(H_1) \approx 30$ dB. From (2.2), we observe that the magnitude response of the synthesis filters are shifted version of the analysis filters and hence are not shown.

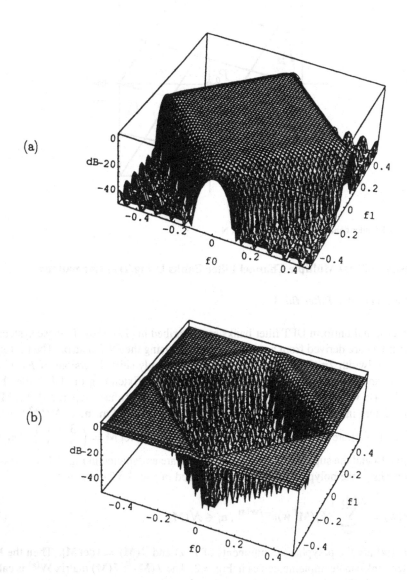

Figure 2.7. Example 2.1. The FIR diamond filter bank. (a) The magnitude response (dB) of $H_0(z)$ and (b) the magnitude response (dB) of $H_1(z)$.

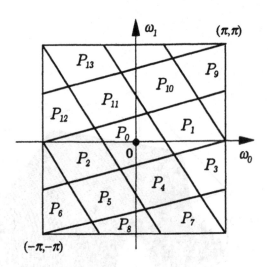

Figure 3.1. DFT filter bank with decimation matrix **N**.

3. Designs of MD Multiple Channel Filter Banks Using Transformations

3.1. Uniform DFT Filter Bank

One-dimensional uniform DFT filter banks are described in [12], [50]. In these systems, a set of M filters are derived from a prototype $P(z)$ by using the DFT matrix. The prototype $P(z)$ has bandwidth $2\pi/M$ and all the filters are uniformly shifted versions of $P(z)$. The shifted amounts are $2\pi k/M$, for $k = 1, 2, \ldots, M - 1$. Extending the DFT filter bank to the MD case with decimation matrix $\mathbf{M}$, the prototype $P(\mathbf{z})$ has support $SPD(\pi\mathbf{M}^{-T})$ and other filters in the filter bank are shifts of $P(\mathbf{z})$ by $2\pi\mathbf{M}^{-T}\mathbf{m}$, $\mathbf{m} \in \mathcal{N}(\mathbf{M}^T)$, where $SPD(\cdot)$ and $\mathcal{N}(\cdot)$ are as defined in Sec. 1.3.1. For example, let $\mathbf{M} = \begin{pmatrix} 3 & -1 \\ 2 & 4 \end{pmatrix}$. This has $|\det \mathbf{M}| = 14$ and the supports of the fourteen filters are as depicted in Fig. 3.1. Expressing $P(\mathbf{z})$ in terms of the polyphase components (defined in Sec. 1.3.2), we have

$$P(\mathbf{w}) = \sum_{i=0}^{J(\mathbf{M})-1} E_i(\mathbf{M}^T\mathbf{w})e^{-j\mathbf{w}^T\mathbf{n}_i}, \, \mathbf{n}_i \in \mathcal{N}(\mathbf{M}), \tag{3.1}$$

where $E_i(\mathbf{w})$ are the polyphase components of $P(\mathbf{z})$ and $J(\mathbf{M}) = |\det \mathbf{M}|$. Then the MD DFT filter bank can be implemented as in Fig. 3.2. The $J(\mathbf{M}) \times J(\mathbf{M})$ matrix $\mathbf{W}^{(g)}$ is called the generalized DFT matrix with elements given by

$$[\mathbf{W}^{(g)}]_{in} = e^{-j2\pi\mathbf{k}_i^T\mathbf{M}^{-1}\mathbf{m}_n}, \, \mathbf{m}_n \in \mathcal{N}(\mathbf{M}), \mathbf{k}_i \in \mathcal{N}(\mathbf{M}^T).$$

As $\mathbf{M}$ is an integer matrix, it admits the decomposition $\mathbf{M} = \mathbf{U}\mathbf{\Lambda}\mathbf{V}$, where $\mathbf{U}$ and $\mathbf{V}$ are unimodular and $\mathbf{\Lambda}$ is a diagonal matrix with $[\mathbf{\Lambda}]_{ii} = \lambda_i$ (see Sec. 1.3.1 for a review of diagonalization of integer matrices). It can be shown that $\mathbf{W}^{(g)}$ assumes the form

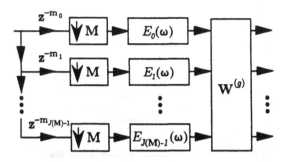

Figure 3.2. The implementation of the MD DFT filter bank, where $\mathbf{W}^{(g)}$ is the generalized DFT matrix.

$$\mathbf{W}^{(g)} = \mathbf{Q}_1 \left(\mathbf{W}_{\lambda_0} \otimes \mathbf{W}_{\lambda_1} \otimes \ldots \otimes \mathbf{W}_{\lambda_{D-1}} \right) \mathbf{Q}_2,$$

where $\mathbf{Q}_1$ and $\mathbf{Q}_2$ are permutation matrices, and $\otimes$ denotes the Kronecker product. The Kronecker product of two matrices $\mathbf{A}$ and $\mathbf{B}$ is defined as

$$\underbrace{\mathbf{A}}_{I \times K} \otimes \underbrace{\mathbf{B}}_{J \times L} = \underbrace{\begin{pmatrix} a_{0,0}\mathbf{B} & \cdots & a_{0,K-1}\mathbf{B} \\ \vdots & \cdots & \vdots \\ a_{I-1,0}\mathbf{B} & \cdots & a_{I-1,K-1}\mathbf{B} \end{pmatrix}}_{IJ \times KL}.$$

When $\mathbf{m}_i$ and $\mathbf{k}_j$ are properly ordered, $\mathbf{Q}_1$ and $\mathbf{Q}_2$ become identity matrices. More specifically, suppose we define these sets of vectors to be

$$\mathbf{m}_n = \mathbf{U}\mathbf{n} \bmod \mathbf{M}, \quad \mathbf{k}_n = \mathbf{V}^T\mathbf{n} \bmod \mathbf{M}^T, \quad \mathbf{n} = (n_0 \ n_1 \ \ldots \ n_{D-1})^T,$$

with subscripts n computed as follows

$$n = n_0 + \lambda_0 n_1 + (\lambda_0 \lambda_1) n_2 + \cdots + (\lambda_0 \lambda_1 \cdots \lambda_{D-2}) n_{D-1}. \tag{3.2}$$

Then we will have $\mathbf{Q}_1 = \mathbf{Q}_2 = \mathbf{I}$. Next, we discuss how to design filters with support $SPD(\mathbf{M}^{-T})$. The method to be introduced below has great design efficiency.

MD filter derived from 1D filter [9]. Recall for a 1D N-fold decimator, the output $Y(\omega)$ is a stretched version of the input $X(\omega)$ by N. For example, consider an input $X(\omega)$ with support $SPD(\pi S)$ (Fig. 3.3(a)). Then the output $Y(\omega)$ will have support $SPD(\pi N S)$ as in (3.3)(b) (assuming no aliasing). Similar situation takes place for MD decimators: the output of an decimation matrix $\mathbf{N}$ is the input "stretched by $\mathbf{N}^T$". That is if the input has support $SPD(\pi \mathbf{S})$ for some matrix $\mathbf{S}$, then the output has support $SPD(\pi \mathbf{N}^T \mathbf{S})$ (assuming no aliasing). We will use this property to design MD filters.

Consider the adjugate of $\mathbf{M}$ defined as $\widehat{\mathbf{M}} \overset{\Delta}{=} J(\mathbf{M})\mathbf{M}^{-1}$. By definition, $\widehat{\mathbf{M}}$ is also an integer matrix. We will see that if the input $X(\mathbf{w})$ of an $\widehat{\mathbf{M}}$-fold decimator has support $SPD(\frac{\pi}{J(\mathbf{M})}\mathbf{I})$, then the output $Y(\mathbf{w})$ has support $SPD(\pi\mathbf{M}^{-T})$. The reason is as follows. Due to $\widehat{\mathbf{M}}$-fold

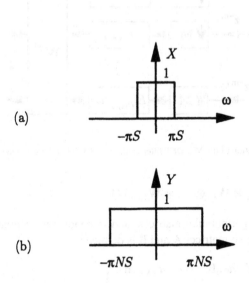

(a)

(b)

Figure 3.3. Relation of the input support and the output support for an N-fold decimator. The output $Y(\omega)$ is a stretched version of the input $X(\omega)$.

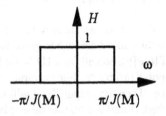

Figure 3.4. The desired response of the 1D filter $H(\omega)$.

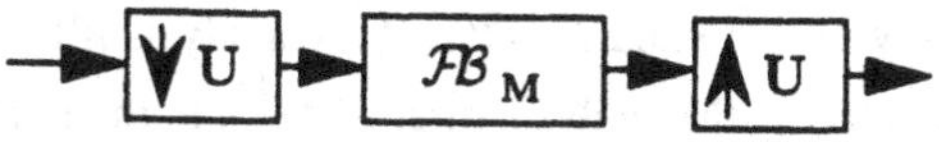

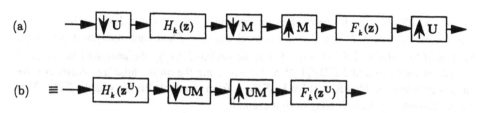

Figure 3.5. Pertaining to the illustration of the unimodular transformation.

Figure 3.6. Derivation of the kth subband of the new system $\mathcal{FB}_{\mathbf{UM}}$.

decimation, $Y(\mathbf{w})$ has support $SPD(\frac{\pi}{J(\mathbf{M})}\widehat{\mathbf{M}}^T)$, which is the same as $SPD(\pi\mathbf{M}^{-T})$ (by the definition of $\widehat{\mathbf{M}}$). So $P(\mathbf{w})$ with support $SPD(\pi\mathbf{M}^{-T})$ can be design through the following procedures.

1. Design a 1D filter $H(\omega)$ with desired response as in Fig. 3.4. Define $h_s(\mathbf{n}) = h(n_0)h(n_1)$ $\cdots h(n_{D-1})$, where $h(n)$ is the impulse response of $H(\omega)$. Then $H_s(\mathbf{w})$ has support $SPD(\frac{\pi}{J(\mathbf{M})}\mathbf{I})$.

2. Let $P(\mathbf{w})$ be the $\widehat{\mathbf{M}}$-fold decimated version of $H_s(\mathbf{w})$, i.e., $p(\mathbf{n}) = h_s(\widehat{\mathbf{M}}\mathbf{n})$. Then $P(\mathbf{w})$ has support $SPD(\pi\mathbf{M}^{-T})$.

Remark. It can be verified that the filters designed through the preceding approach have separable polyphase components. In this case, if $E_i(\mathbf{w})$ are the polyphase components of $P(\mathbf{w})$ as in (3.1), then $E_i(\mathbf{w})$ is separable and can be expressed as the product of polyphase components of the 1D filter $h(n)$. So the complexity of the DFT filter bank implementation in Fig. 3.2 grows linearly with the number of dimensions.

3.2. Unimodular Transformation [49]

Consider the D-dimensional filter bank with integer decimation matrix $\mathbf{M}$ in Fig. 1.1. This filter bank will be denoted by $\mathcal{FB}_{\mathbf{M}}$. Suppose the filter bank has perfect reconstruction. Recall that a $D \times D$ unimodular decimator $\mathbf{U}$ or a unimodular expander $\mathbf{U}$ only permutes the input. So if we insert a decimator $\mathbf{U}$ before the filter bank $\mathcal{FB}_{\mathbf{M}}$ and an expander $\mathbf{U}$ after $\mathcal{FB}_{\mathbf{M}}$ (Fig. 3.5), the new system still has perfect reconstruction. This is equivalent to inserting a decimator $\mathbf{U}$ before the analysis filter and an expander $\mathbf{U}$ after the synthesis filter in each subband (Fig. 3.6(a)). We can redraw Fig. 3.6(a) as Fig. 3.6(b). Denote the new filter bank with decimation matrix $\mathbf{UM}$ by $\mathcal{FB}_{\mathbf{UM}}$. The system $\mathcal{FB}_{\mathbf{UM}}$ will be called the unimodular transformation of $\mathcal{FB}_{\mathbf{M}}$ by $\mathbf{U}$. The kth analysis filter and the kth synthesis filter of $\mathcal{FB}_{\mathbf{UM}}$ are respectively $H_k(\mathbf{z}^{\mathbf{U}})$ and $F_k(\mathbf{z}^{\mathbf{U}})$. If the analysis filter $H_k(\mathbf{z})$ in the original filter bank $\mathcal{FB}_{\mathbf{M}}$ has support $\mathcal{S}_k$, the analysis filter $H_k(\mathbf{z}^{\mathbf{U}})$ in the new system $\mathcal{FB}_{\mathbf{UM}}$ has support $\mathbf{U}^{-T}\mathcal{S}_k$ given by

$$\mathbf{U}^{-T}\mathcal{S}_k = \{\mathbf{U}^{-T}\mathbf{w}, \mathbf{w} \in \mathcal{S}_k\}.$$

Now consider the special case when the original filter bank is a 2D separable system with $\mathbf{M} = \Lambda$, where Λ is a 2×2 diagonal matrix. Then each analysis filter $H_k(\mathbf{z})$ consists of four shifts of $SPD(\frac{\pi}{2}\Lambda^{-1})$. For the system $\mathcal{FB}_{\mathbf{U}\Lambda}$, the analysis filter $H_k(\mathbf{z}^{\mathbf{U}})$ consists of four shifts of $SPD(\frac{\pi}{2}\mathbf{U}^{-T}\Lambda^{-1})$. So, using the unimodular transformation we can design nonseparable PR filter banks $\mathcal{FB}_{\mathbf{U}\Lambda}$ from separable PR filter banks $\mathcal{FB}_{\Lambda}$. This is demonstrated by the following design example.

Example 3.1. Unimodular transformation. Let $\Lambda = \begin{pmatrix} 4 & 0 \\ 0 & 5 \end{pmatrix}$. We can design a separable 20-channel filter bank $\mathcal{FB}_{\Lambda}$ with decimation matrix Λ by concatenating a 1D four-channel filter bank and a 1D five-channel filter bank. Let $\mathcal{FB}_{\mathbf{U}\Lambda}$ be the unimodular transformation of $\mathcal{FB}_{\Lambda}$ by $\mathbf{U}$, where $\mathbf{U} = \begin{pmatrix} 2 & -1 \\ -1 & 1 \end{pmatrix}$, then $\mathcal{FB}_{\mathbf{U}\Lambda}$ is nonseparable. Supports of some of the analysis filters in $\mathcal{FB}_{\mathbf{U}\Lambda}$ are shown in Fig. 3.7(a). Fig. 3.7(b) shows the magnitude response of the lowpass analysis filter.

3.3. Unimodular Transformation and Two-Channel Filter Banks

Recall the two-channel filter banks with parallelogram-supported filters as in Fig. 1.7. It can be verified that the supports of the lowpass analysis or synthesis filters in those four cases of Fig. 1.7 are of the form $SPD(\pi\mathbf{M}_i^{-T})$, with $\mathbf{M}_i = \mathbf{U}_i\mathbf{Q}$, where $\mathbf{U}_i$ are unimodular and $\mathbf{Q}$ is the quincunx matrix in Fig. 1.2. The matrix $\mathbf{U}_i$ for the cases (a), (b), (c) and (d) are respectively

$$\mathbf{U}_1 = \begin{pmatrix} 1 & -1 \\ 0 & 1 \end{pmatrix}, \mathbf{U}_2 = \begin{pmatrix} -1 & 1 \\ 0 & 1, \end{pmatrix} \mathbf{U}_3 = \begin{pmatrix} 0 & -1 \\ 1 & -1 \end{pmatrix} \quad \mathbf{U}_4 = \begin{pmatrix} 0 & 1 \\ 1 & -1. \end{pmatrix}.$$

We can choose the decimation matrix to be $\mathbf{M}_i$. Since $\mathbf{M}_i = \mathbf{U}_i\mathbf{Q}$, where $\mathbf{U}_i$ are unimodular, these filter banks can be derived from the diamond filter bank. Thus, these systems are unimodular transformations of the diamond filter bank.

It can shown that any 2×2 integer matrix $\mathbf{M}$ with $|\det\mathbf{M}| = 2$ can be expressed as one of the following three forms

$$\mathbf{M} = \mathbf{U}\begin{pmatrix} 1 & 0 \\ 0 & 2, \end{pmatrix} \mathbf{M} = \mathbf{U}\begin{pmatrix} 2 & 0 \\ 0 & 1, \end{pmatrix} \text{ or } \mathbf{M} = \mathbf{U}\mathbf{Q},$$

where $\mathbf{Q}$ is the quincunx matrix. So the corresponding two-channel filter bank with parallelogram supports can always be obtained as the unimodular transformation of the diamond filter bank or the unimodular transformation of 1D filter banks. In the quadrant filter bank

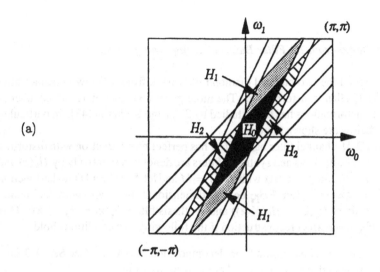

(a)

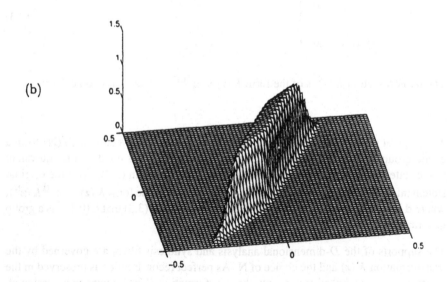

(b)

Figure 3.7. Example 3.1. Unimodular transformation. (a) Spectral supports of some of the analysis filters and (b) the magnitude response of the low-pass analysis filter with frequency normalized by 2π.

(Fig. 1.5), supports of the analysis filters are not parallelograms; the quadrant filter bank can not be obtained as the unimodular transformation of the diamond filter bank or 1D filter banks.

3.4. MD Filter Banks from 1D Filter Banks by using Transformation

In Sec. II, we designed the two-channel diamond filter bank from a 1D two-channel filter bank by use of McClellan transformation. The more general subject of 1D M-channel to MD M-channel transformations has been studied by Shah and Kalker in [43]. In particular, the results given below are shown therein.

Suppose that a 1D M-channel filter bank $\mathcal{FB}_M$ has perfect reconstruction with distortion function $T(z) = 1$. The analysis and synthesis filters are denoted respectively by $H_k(z)$ and $F_k(z)$. Let N be a $D \times D$ integer matrix with $J(N) = M$ and let $K(z)$ be a 1D scalar function of $\mathbf{z} = (z_0 \ z_1 \ \cdots \ z_{D-1})^T$. Let $\mathcal{FB}_N$ be a D-dimensional filter bank with decimation matrix N, analysis filters $H'_k(\mathbf{z}) = H_k(K(\mathbf{z}))$ and synthesis filters $F'_k(\mathbf{z}) = F_k(K(\mathbf{z}))$. One can show that $\mathcal{FB}_N$ has perfect reconstruction if the following two conditions hold.

1. The decimation matrix N has Smith form decomposition $N = U\Lambda V$ (see Sec. 1.3 for a review of Smith form decomposition), where Λ is of the form

$$\Lambda = \begin{pmatrix} 1 & 0 & & 0 \\ 0 & 1 & & 0 \\ \vdots & & \ddots & \vdots \\ 0 & 0 & & M \end{pmatrix}. \tag{3.3}$$

2. The transformation $K(\mathbf{z})$ is of the form $K(\mathbf{z}) = \mathbf{z}^{-U(0\ 1)^T} \widehat{K}(\mathbf{z}^N)$, for some $\widehat{K}(\mathbf{z})$.

Remarks.

1. The original statement of the first condition in [43] is that the vectors in $\mathcal{N}(N)$ form a cyclic group under modulo N. However as shown in [8], this condition is equivalent to the statement that N has the special Smith form as given in (3.3). Also the original statement of the second condition in [43] is that $K(\mathbf{z})$ is of the form $K(\mathbf{z}) = \mathbf{z}^{-\mathbf{d}} \widehat{K}(\mathbf{z}^N)$, where $\mathbf{d}$ is a group generator of $\mathcal{N}(N)$. But it follows from (3.3) that $U(0\ 1)^T$ is a group generator of $\mathcal{N}(N)$.

2. The supports of the D-dimensional analysis and synthesis filters are governed by the transformation $K(\mathbf{z})$ and the choice of N. As perfect reconstruction is preserved in the transformation described above, only the transformation $K(\mathbf{z})$ remains to be designed. For the filter bank with diamond filters, the McClellan transformation is appropriate for realizing the 2D diamond filters. However, except in this case, there is no systematic approach of finding the transformation $K(\mathbf{z})$ that controls the shape of the 2D (more generally, D dimensional) analysis and synthesis filters. In general, the transformation $K(\mathbf{z})$ does not faithfully translate the characteristics of 1D filters. So even if the

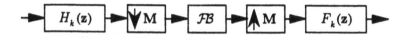

Figure 4.1. Derivation of perfect reconstruction tree structured filter bank.

1D analysis and synthesis filters in $\mathcal{FB}_M$ have good stopband attenuation and good passbands, the D-dimensional analysis and synthesis filters resulting from the previous formulation may not have these properties.

4. Tree Structured Filter Banks

Consider the filter bank in Fig. 1.1. In the kth subband, we can insert another filter bank $\mathcal{FB}$ (Fig. 4.1) to obtain a tree-structured filter bank. If the filter bank in Fig. 1.1 has perfect reconstruction (PR) and $\mathcal{FB}$ is also PR with distortion function $T(\mathbf{z}) = 1$, the overall system remains PR. In this case, we say that the filter bank in Fig. 1.1 is the first level of the tree and $\mathcal{FB}$ is the second level. Repeat this insertion of PR filter bank, we can obtain a tree structure of many levels. The separable systems are examples of tree-structured filter banks in which all the member filter banks are one-dimensional. In 2D, tree structure of some extensively-studied systems, e.g. the two-channel filter banks or the separable filter banks, sometimes offers very sophisticated supports.

Example 4.1. Directional filter banks [3]. The directional filter bank with supports as in Fig. 1.9 can be obtained by using a tree structure to cascade some commonly used two-channel systems. In particular, it can be obtained by cascading filter banks with fan filters (Fig. 1.6) and filter banks with supports as in Fig. 1.7 (except some minor modifications). We will explain in detail how to obtain $H_0(\mathbf{z})$ and $H_1(\mathbf{z})$ in Fig. 1.9. The supports of other analysis filters can be obtained in a similar manner. Consider the tree structure in Fig. 4.2. Let the analysis filters $G_0(\mathbf{z})$ and $G_1(\mathbf{z})$ have fan supports as in Fig. 1.6. Also let the support of $K_0(\mathbf{z})$ and $K_1(\mathbf{z})$ be as in Fig. 4.3, which are shifts of the parallelograms in Fig. 1.7(a) by $(\pi\ \pi)^T$. By use of Noble identities (Sec. 1.3), Fig. 4.2 can be redrawn as Fig. 4.4. The filters $G_0(\mathbf{z}^{\mathbf{Q}})$, $K_0(\mathbf{z}^{2\mathbf{I}_2})$ and $K_1(\mathbf{z}^{2\mathbf{I}_2})$ respectively have supports shown in Fig. 4.5. So $G_0(\mathbf{z})G_0(\mathbf{z}^{\mathbf{Q}})K_0(\mathbf{z}^{2\mathbf{I}_2})$ and $G_0(\mathbf{z})G_0(\mathbf{z}^{\mathbf{Q}})K_1(\mathbf{z}^{2\mathbf{I}_2})$ yield the desired support for $H_0(\mathbf{z})$ and $H_1(\mathbf{z})$.

Example 4.2. Consider the hexagon matrix $\mathbf{M} = \begin{pmatrix} 1 & 1 \\ -2 & 2 \end{pmatrix}$, which can be factorized as

$$\mathbf{M} = \begin{pmatrix} 1 & 0 \\ 0 & 2 \end{pmatrix} \begin{pmatrix} 1 & 1 \\ -1 & 1 \end{pmatrix}.$$

Suppose the first level of the tree is a separable system as in Fig. 1.3(b) and the second level is the diamond filter bank (Fig. 1.4). Then the four analysis filters of the overall system have supports as shown in Fig. 4.6.

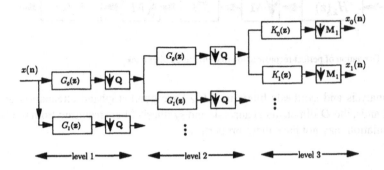

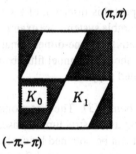

Figure 4.2. Tree-structured filter bank for the derivation of directional filter banks. In the first and second levels, each subband is split into two. Only the first two subbands of the overall system are shown in the figure.

Figure 4.3. The support of the filters $K_0(z)$ and $K_1(z)$.

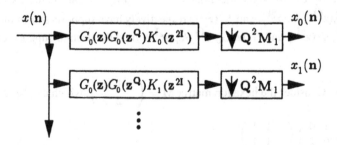

Figure 4.4. The equivalent parallel structure of Fig. 4.2.

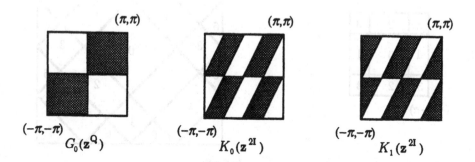

Figure 4.5. The supports of $G_0(z^Q)$, $K_0(z^{2I_2})$ and $K_1(z^{2I_2})$.

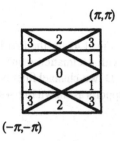

Figure 4.6. Example 4.2. Tree structured filter bank. The supports of the analysis filters. In the figure the support of $H_k(z)$ is denoted by k.

39

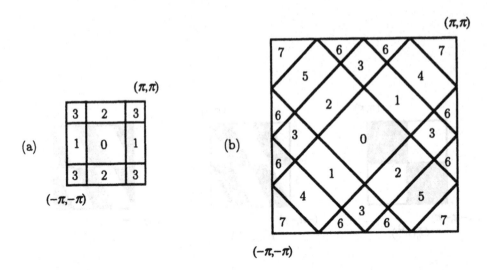

Figure 4.7. Example 4.3. Tree structured filter bank. (a) The first level of the tree: the four-channel separable filter bank. The supports of the four separable analysis filters (with the support of the *k*th filter denoted by *k*). (b) The supports of the eight analysis filters in the overall system. The support of $H_k(z)$ is denoted by *k*.

Example 4.3. Let

$$\mathbf{M} = \begin{pmatrix} 1 & 1 \\ -1 & 1 \end{pmatrix} \begin{pmatrix} 2 & 0 \\ 0 & 2 \end{pmatrix}.$$

If we use the diamond filter bank (Fig. 1.4) for the first level of the tree. For the second level, we use a separable system with decimation matrix $2\mathbf{I}_2$ (Fig. 4.7(a)), which can be obtained as a tree structure of two 1D two-channel filter banks. Then the resulting analysis filters of the overall system have supports as shown Fig. 4.7(b). Each analysis filter consists of four parallelograms.

5. Two-Dimensional Cosine Modulated Filter Banks: Basic Consideration

We first recall the main features of one-dimensional (1D) cosine modulated filter banks (CMFB). The process of constructing 1D CMFB will give us a general idea of the procedures for constructing 2D CMFB. The experience with 1D CMFB will help also us gain some insight to foresee the difficulties in constructing the 2D CMFB.

5.1. One-Dimensional Cosine Modulated Filter Banks (CMFB)

5.1.1. Introduction

Two types of cosine modulated filter banks have been developed, pseudo QMF systems [10], [41], [37] and perfect reconstruction systems [39], [35], [26]. Consider the filter bank in Fig. 1.1, in which the decimation matrix is a scalar M. An M-channel CMFB (pseudo-QMF or perfect reconstruction) is typically obtained by starting from a $2M$ channel uniform DFT filter bank, [50]. Each filter in the DFT filter bank is a shifted version of a lowpass prototype $P(\omega)$ (Fig. 5.1(a)) with bandwidth π/M, which is half the total bandwidth of each filter in the desired M-channel system. The filters in the DFT filter bank are then shifted by $\pi/2M$ and paired to obtain real-coefficient analysis filters as in Fig. 5.1(b) for the M channel CMFB. The shifts of the prototype are denoted by $P_k(\omega)$ in the figure. The total bandwidth of each of the analysis filters in the CMFB is $2\pi/M$, which is two times that of the prototype. In almost all the designs, the synthesis filters are time-reversed versions of the corresponding analysis filters; the analysis and synthesis filters have the same spectral support.

In the CMFB described above, as each analysis filter consists of two shifted copies of the prototype, each of the two copies has $M-1$ images due to decimation followed by expansion. By construction, the images of the analysis filters are adjacent to the support of the corresponding synthesis filters but are not overlapping with the passbands of synthesis filters as shown in Fig. 5.1(c). Thus, if the prototype filter is an ideal brick-wall filter, there is no aliasing and in this case the filter bank has perfect reconstruction. If the prototype filter is not ideal, those images that are adjacent to the synthesis filter result in major aliasing (Fig. 5.1(d)) while those that are not adjacent to the synthesis filters will be attenuated to the stopband level of the prototype filter. In the pseudo QMF CMFB, only the major aliasing errors are canceled and approximate alias cancellation is attained. Approximate reconstruction is then achieved without sophisticated optimization of the lowpass prototype. In the prefect reconstruction CMFB, the prototype is further optimized under the constraint that the CMFB is paraunitary, hence perfect reconstruction is assured. The paraunitariness of the CMFB is guaranteed if the polyphase components of the prototype filter satisfy some pairwise power complementary conditions [26]. In both pseudo QMF and perfect reconstruction systems, the design of the whole filter bank is reduced to the optimization of the lowpass prototype filter. The complexity of the analysis bank is equal to that of a prototype filter plus a DCT matrix.

5.1.2. General setting and main features of 1D CMFB

From the above discussion of 1D CMFB, we observe that the general setting of 1D M-channel CMFB can be summarized as follows.

1. Construct a $2M$-channel uniform DFT filter bank.

2. Shift the filters in the DFT filter bank by $\pi/2M$ and combine appropriate pairs of filters to yield real-coefficient analysis filters.

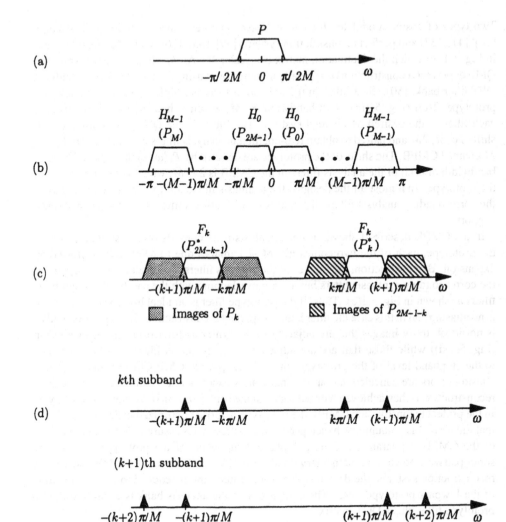

Figure 5.1. One-dimensional cosine modulated filter bank. (a) The support of the prototype filter $P(\omega)$. (b) The support of the analysis filters $H_k(\omega)$. Each analysis filter has two parts, $P_k(\omega)$ and $P_{2M-1-k}(\omega)$. (c) Images of the analysis filter $H_k(\omega)$ that are adjacent to the synthesis filter $F_k(\omega)$. (d) The major aliasing in the kth subband and the $(k+1)$th subband.

Main features of 1D CMFB. The support configuration of the 1D CMFB constructed above has the following two features. (1) The support configuration satisfies the **bandpass sampling criterion**: images of the analysis filter passbands do not overlap with passbands of the synthesis filters. Furthermore, the filter bank has perfect reconstruction when the analysis and synthesis filters are ideal brick-wall filters. (2) When the filters are not ideal, the major aliasing errors that contribute to the same aliasing transfer function $A_i(\mathbf{w})$ appear in pairs. For example, both kth and $(k+1)$th subbands have major aliasing errors around the frequency $k\pi/M$ (Fig. 5.1(d)) and it can be verified that these two aliasing errors contribute to the same aliasing transfer function $A_k(\mathbf{w})$. A support configuration without this feature will be referred to as *nonpermissible*. The importance of satisfying the sampling criterion and the significance of support permissibility are addressed next.

Satisfying the sampling criterion. This feature is indispensable for the design of perfect reconstruction filter banks. For a configuration that does not satisfy the sampling criterion, severe aliasing will be created in the subbands no matter how good the filters are. A filter bank with a configuration that violates the sampling criterion can not have perfect reconstruction even if the analysis and synthesis filters are ideal brick-wall filters.

Support permissibility. Support permissibility allows the possibility of cancellation of major aliasing. Pairwise major aliasing terms are necessary if cancellation of major aliasing is to take place. For a filter bank with a nonpermissible support configuration, the filters can not have good stopband attenuation [8]. The reason is as follows. Suppose a perfect reconstruction filter bank has a nonpermissible support configuration and the individual filters have good stopband attenuation. Then, major aliasing terms do not appear in pairs and can not be cancelled. This contradicts the fact that the filter bank has perfect reconstruction. Therefore, a PR filter bank with nonpermissible support configuration can not have filters with good stopband attenuation except in the case that all the filters are ideal brick-wall filters.

Example of nonpermissible support. The configuration of the DFT filter bank is an example of nonpermissible support. The DFT filter bank has the first feature of 1D CMFB (i.e. satisfying the bandpass sampling criterion) but not the second one. As a result, the individual analysis and synthesis filters in a perfect reconstruction DFT filter bank can not have good attenuation unless all the filters are ideal brick-wall filters. To see this, consider a three-channel DFT filter bank. The supports of the analysis filter are as shown in Fig. 5.2(a). The subband signals are decimated by 3 and expanded by 3, so each analysis filter has two images. For example, the images of $H_0(z)$ are as shown in Fig. 5.2(b). The analysis filter $H_k(z)$ and its images do not overlap in the passbands. If the prototype filter is ideal, no aliasing is created in the subbands and the filter bank has perfect reconstruction. So the support configuration of the DFT filter bank satisfies the sampling criterion and has the first feature of the CMFB. When the filters have good stopband attenuation, the major aliasing errors result from the two images of $H_0(z)$ are shown in Fig. 5.2(c). The aliasing around the frequency $\pi/3$ contribute to the alias transfer function $A_1(z)$ while the aliasing error around the frequency $-\pi/3$ contribute to the alias transfer function $A_2(z)$. Similar aliasing errors occur in the other two subbands as well (Fig. 5.2(c)). We observe that around the frequency $\pi/3$, both the first and the second subband have aliasing errors. However, these two aliasing errors contribute to different aliasing transfer functions and can not cancel

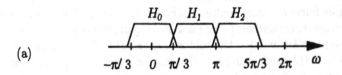

(a)

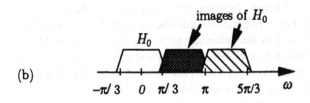

(b)

(c)   first subband

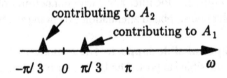

second subband

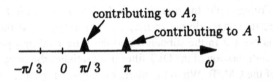

Figure 5.2. Three-channel DFT filter bank. (a) The supports of the three analysis filters. (b) Images of $H_0(\omega)$ due to 3-fold decimation followed by 3-fold expansion. (c) Major aliasing in the first and second subbands.

each other. So for the DFT filter bank, aliasing can not be cancelled if the individual filters have good attenuation but are not ideal filters. In a perfect reconstruction filter bank, aliasing eventually has to be cancelled. We hence conclude that the filters in the perfect reconstruction DFT filter bank can not have good stopband attenuation except in the case when the filters are ideal.

5.2. Construction of Two-Dimensional Cosine Modulated Filter Bank

The separable 2D CMFB can always be obtained through concatenation of two 1D CMFB in the form of a tree [50]. Our interest in this paper is designing a nonseparable 2D CMFB. The prototype filter is in general a nonseparable 2D filter. Each analysis and synthesis filter is a cosine modulated version of the prototype and is also nonseparable. In the separable 2D CMFB case, each individual filter consists of four shifts of a separable 2D prototype. However, the real-coefficient constraint on the analysis filters requires only two copies of the prototype. We can therefore conceive that in the more general case the individual filters can have two or four shifted copies of the prototype. In this paper, we study two classes of 2D FIR paraunitary cosine modulated filter banks: the two-copy CMFB and the four-copy CMFB. In the two-copy CMFB, each individual filter contains two shifted copies of the prototype and in the four-copy CMFB, each individual filter contains four shifted copies of the prototype. The filter bank will eventually be constrained to be paraunitary; the analysis filter $H_k(\mathbf{w})$ and the corresponding synthesis filter $F_k(\mathbf{w})$ are then related by $F_k(\mathbf{w}) = H_k^*(\mathbf{w})$ for perfect reconstruction. The analysis and synthesis filters have the same spectral support. Note that the two-copy CMFB is fundamentally different from a 2D separable CMFB obtained from two 1D CMFB. But the four-copy CMFB will reduce to separable 2D CMFB in special cases. Both of these systems are elaborated below.

Two-copy CMFB. Consider the filter bank in Fig. 1.1 with decimation matrix $\mathbf{M}$, non-diagonal in general. There are $|\det \mathbf{M}|$ channels since each decimator creates a decimation of $|\det \mathbf{M}|$. For a given $|\det \mathbf{M}|$-channel filter bank as in Fig. 1.1, we start from a $2|\det \mathbf{M}|$-channel uniform 2D DFT filter bank [50]. Every filter in the DFT filter bank is a shifted version of a lowpass FIR (non-rectangular) prototype. The prototype has a parallelogram support, so every filter in the DFT filter bank has a parallelogram support. The filters in the DFT filter bank are shifted and paired to obtain real-coefficient analysis filters. Each analysis filter is then a cosine modulated version of the prototype and each analysis filter consists of two shifted copies of the prototype. All the analysis and synthesis filters have real coefficients. We then examine whether the supports of the analysis and synthesis filters satisfy the sampling criterion. Namely, the images of the analysis filter passbands should not overlap with passbands of the synthesis filters. Under this condition we study the sufficient conditions such that cancellation of major aliasing (due to overlapping transition bands) can be structurally enforced. Finally, having cancelled the major aliasing, we constrain the prototype to ensure perfect reconstruction of the two-copy CMFB.

Four-copy CMFB. The four-copy CMFB will be constructed in a similar way. But in the four-copy case, we design a $4|\det \mathbf{M}|$-channel uniform DFT filter bank. Then we shift the filters in the DFT filter bank and combine four shifted filters to obtain real-coefficient analysis filters for the four-copy CMFB. The rest of the construction procedure is the same as that of two-copy CMFB.

The above construction of two-copy and four-copy CMFB is an immediate imitation of 1D CMFB. The more refined construction procedures are listed below.

1. **General setting.** To complete the general setting of the 2D CMFB, we need to the answer the following questions first. For a given filter bank with decimation matrix $\mathbf{M}$ as in Fig. 1.1, how to construct the uniform DFT filter bank? How to shift the filters in the DFT filter bank and obtain the analysis filters such that the support configuration of the 2D CMFB is an extension of 1D version? In 1D case, there is only one way to shift the filters of the DFT filter bank. Do we have more variety in 2D case?

2. **Ideal case.** Having completed the general setting of 2D CMFB, we proceed to examine the support configuration of 2D CMFB. We first check whether the configuration has the first feature of 1D CMFB. The configuration should be such that the bandpass sampling criterion is satisfied and the filter bank has perfect reconstruction in the case of ideal filters. To satisfy the sampling criterion, the images of passbands of an analysis filter should not overlap with passbands of the corresponding synthesis filter.

3. **Support permissibility and alias cancellation.** For those that satisfy the sampling criterion, we examine support permissibility. Permissibility of a support configuration means that cancellation of major aliasing is possible. If a support configuration is permissible, we further study how to cancel major aliasing.

4. **Perfect reconstruction.** Having cancelled major aliasing, we then constrain the prototype to achieve perfect reconstruction as in the case of 1D CMFB.

6. Two-Copy Cosine Modulated Filter Banks

6.1. General Setting of Two-Copy CMFB

1. **Uniform DFT filter bank.** To design a two-copy CMFB with decimation matrix $\mathbf{M}$, we start from a uniform DFT filter bank [50] with decimation matrix $\mathbf{N} = \mathbf{ML}$, where $\mathbf{L}$ is an integer matrix (to be chosen appropriately) with $|\det \mathbf{L}| = 2$. For example, let

$$\mathbf{M} = \begin{pmatrix} 7 & -2 \\ 0 & 1 \end{pmatrix} \quad \text{and} \quad \mathbf{L} = \begin{pmatrix} 1 & 1 \\ 2 & 4 \end{pmatrix}, \quad \text{then} \quad \mathbf{N} = \begin{pmatrix} 3 & -1 \\ 2 & 4 \end{pmatrix}.$$

The supports of the individual filters $\{P_k(\mathbf{w}), 0 \le k \le 13\}$ in the DFT filter bank are as shown in Fig. 3.1. Each filter in the DFT filter bank is a shifted version of the prototype $P(\mathbf{w})$, which has a parallelogram support $SPD(\pi \mathbf{N}^{-T})$. For a given $\mathbf{M}$, the support of the prototype is different for different choice of $\mathbf{L}$.

2. **The analysis and synthesis filters.** In 1D CMFB, we shift the filters in the DFT filter bank by $\pi/2M$. But in 2D case, the shifts are vector-shifts. We can shift the filters in the DFT filter bank in three possible directions as in Fig. 6.1. The result supports with respect to the three different shifts are shown in Fig. 6.2(a)–(c). For all the three cases, filters can be paired to obtain real-coefficient analysis filters for the two-copy CMFB. For example, in Fig. 6.2(a), the filter coefficients of $Q_{A,i}(\mathbf{w})$ and $Q'_{A,i}(\mathbf{w})$ are complex conjugates of each

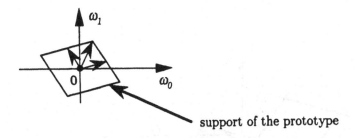

Figure 6.1. Three possible vector shifts.

other and can be paired to obtain the analysis filter,

$$H_{A,i}(\mathbf{w}) = Q_{A,i}(\mathbf{w}) + Q'_{A,i}(\mathbf{w}). \tag{6.1}$$

The corresponding synthesis filter is

$$F_{A,i}(\mathbf{w}) = Q^*_{A,i}(\mathbf{w}) + Q'^*_{A,i}(\mathbf{w}).$$

Similarly in Fig. 6.2(b), $Q_{B,i}(\mathbf{w})$ and $Q'_{B,i}(\mathbf{w})$ are combined to obtain the analysis filter $H_{B,i}(\mathbf{w})$ and in Fig. 6.2(c) $Q_{C,i}(\mathbf{w})$ and $Q'_{C,i}(\mathbf{w})$ are combined to obtain $H_{C,i}(\mathbf{w})$. Each analysis filter consists of two parallelograms. We observe that all three possible support configurations are extension of the 1D version. The three support configurations will be referred to as configurations A, B, and C. From Fig. 6.2 it seems that configurations A and B are very similar. Indeed as we will see in subsequent discussion, properties derived for configurations A are also true for configuration B except some minor modifications.

6.2. The Ideal Case

In the above general setting of two-copy CMFB, for a given decimation matrix $\mathbf{M}$, we first construct a DFT filter bank with decimation matrix $\mathbf{N} = \mathbf{ML}$. The matrix $\mathbf{L}$ is an integer matrix with determinant 2. No additional assumption has been made on $\mathbf{L}$. On the other hand, given a DFT filter bank with decimation matrix $\mathbf{N}$, we can always shift the filters in the DFT filter bank by the three different amounts indicated in Fig. 6.1. These three shifts gives us configurations A, B, and C. So the derivation of configurations A, B, and C does not relate to $\mathbf{L}$. However, we will see that for a given $\mathbf{L}$, not all three configurations can satisfy the bandpass sampling criterion. In particular, we will establish a connection between the lattice of $\mathbf{L}^T$ and valid choice of configurations.

As $\mathbf{L}$ has determinant 2, there are only three possible choices for $LAT(\mathbf{L}^T)$ as in Fig. 6.3(a)–(c). The lattice of $\mathbf{L}^T$ is rectangular in Fig. 6.3(a),(b) and quincunx in Fig. 6.3(c). We will argue that for each configuration, there are only two choices of $LAT(\mathbf{L}^T)$ that can satisfy the sampling criterion and these combinations are the valid candidates for the development of two-copy CMFB. The relation between the three configurations and valid choice of $LAT(\mathbf{L}^T)$ is summarized in Table 6.1. In this case, when the sampling criterion

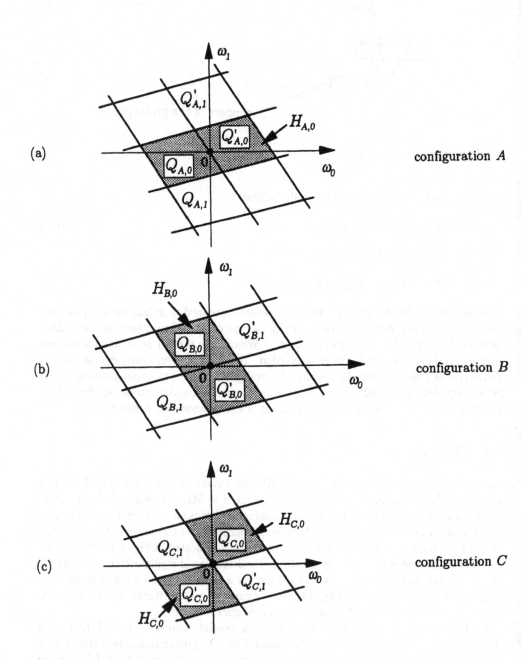

(a)   configuration A

(b)   configuration B

(c)   configuration C

Figure 6.2. Three possible support configurations.

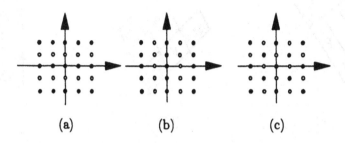

(a)   (b)   (c)

o Integers

• Integers on lattice of $\mathbf{L}^T$

Figure 6.3. Possible lattices generated by $\mathbf{L}^T$, when $|\det \mathbf{L}| = 2$.

$LAT(\mathbf{L}^T)$ configuration	lattice (a) (rectangular)	lattice (b) (rectangular)	lattice (c) (quincunx)
A		satisfied	satisfied
B	satisfied		satisfied
C	satisfied	satisfied	

Table 6.1. Possible combinations of configurations and $LAT(\mathbf{L}^T)$ that satisfy the sampling criterion. Lattice (a)–(c) in the table are as in Fig. 6.3 (a)–(c).

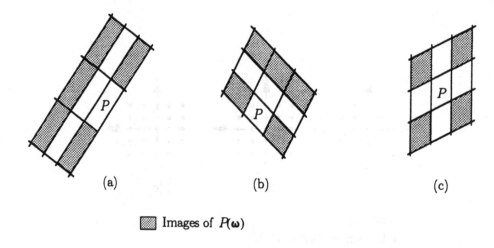

(a)   (b)   (c)

▨ Images of $P(\omega)$

Figure 6.4. Images of the prototype $P(\mathbf{w})$ with respect to the three possible lattices generated by $\mathbf{L}^T$.

is satisfied, the filter bank has perfect reconstruction if the prototype is an ideal brick-wall filter. The remaining part of this subsection is devoted to the verification of Table 6.1.

To check whether a certain configuration satisfies the sampling criterion, it is essential to know exactly where the images of the analysis filters are located. As each analysis filter contains two shifted copies of the prototype, we first inspect the locations of images of $P(\mathbf{w})$ when it is decimated by $\mathbf{M}$ and then expanded by $\mathbf{M}$. The images of $P(\mathbf{w})$ are [50]

$$P(\mathbf{w} - 2\pi\mathbf{M}^{-T}\mathbf{k}), \quad \mathbf{k} \in \mathcal{N}(\mathbf{M}^T) \quad \text{and} \quad \mathbf{k} \neq \underline{0}.$$

Recall that the support of $P(\mathbf{w})$ is $SPD(\pi\mathbf{N}^{-T})$, which depends on both $\mathbf{M}$ and $\mathbf{L}$. For different $LAT(\mathbf{L}^T)$, $P(\mathbf{w})$ and its images form a certain pattern. In particular, it is a striped pattern in the case of rectangular $LAT(\mathbf{L}^T)$ and a check pattern in the case of quincunx $LAT(\mathbf{L}^T)$. For example, let $\mathbf{M} = \begin{pmatrix} 2 & 1 \\ -1 & 1 \end{pmatrix}$. When

$$\mathbf{L} = \begin{pmatrix} 1 & 0 \\ 0 & 2 \end{pmatrix}, \quad \mathbf{L} = \begin{pmatrix} 2 & 0 \\ 0 & 1 \end{pmatrix}, \quad \text{and} \quad \mathbf{L} = \begin{pmatrix} 1 & -1 \\ 1 & 1 \end{pmatrix},$$

the locations of the images are shown respectively in Fig. 6.4(a)–(c). In all three cases, images of $P(\mathbf{w})$ are confined to the grid formed by shifts of $P(\mathbf{w})$ in the DFT filter bank.

The lattice of $\mathbf{L}^T$ and the sampling criterion

Configuration A. In what follows configuration A of Fig. 6.2(a) will be shown to satisfy the sampling criterion when $LAT(\mathbf{L}^T)$ is as in Fig. 6.3(b) and (c). For these two types of $LAT(\mathbf{L}^T)$, the filter bank has perfect reconstruction when the prototype is an ideal brick-wall filter. Consider the lowpass filter $H_{A,0}(\mathbf{w})$. When $LAT(\mathbf{L}^T)$ is rectangular as in

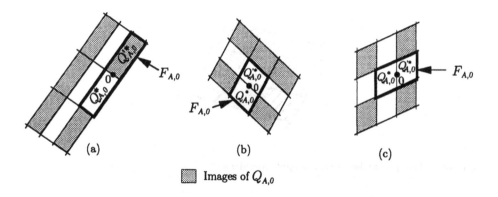

(a)   (b)   (c)

Images of $Q_{A,0}$

Figure 6.5. Configuration A and the sampling criterion. Images of $Q_{A,0}(\mathbf{w})$ with respect to the three different lattices generated by $\mathbf{L}^T$.

Fig. 6.3(a), the support of $Q_{A,0}(\mathbf{w})$ and its images (shaded areas) are shown in Fig. 6.5(a). One image of $Q_{A,0}(\mathbf{w})$ completely falls on top of $Q'^*_{A,0}(\mathbf{w})$, which is part of the synthesis filter $F_{A,0}(\mathbf{w})$; the sampling criterion is not satisfied. When lattice of $\mathbf{L}^T$ is as in Fig. 6.3(b) or (c), the images of $Q_{A,0}(\mathbf{w})$ are shown in Fig. 6.5(b) and (c). In these two cases the images of $Q_{A,0}(\mathbf{w})$ are edge-adjacent to the synthesis filter $F_{A,0}(\mathbf{w})$ but not overlapping with $F_{A,0}(\mathbf{w})$. It can be verified that in every other subband, similar situation takes place: images of the analysis filters are adjacent to the synthesis filters but not overlapping with support of synthesis filters. Therefore, for the lattices of $\mathbf{L}^T$ shown in Fig. 6.3(b), (c), configuration A satisfies the sampling criterion.

Configuration B and C. Following a similar analysis, configuration B can be shown to satisfy the sampling criterion when $LAT(\mathbf{L}^T)$ is as given in Fig. 6.3(a) and (c). Also configuration C can be shown to satisfy the sampling criterion when $LAT(\mathbf{L}^T)$ is rectangular as given in Fig. 6.3(a) and (b). Moreover, in all these cases (configuration B or C) the filter bank has perfect reconstruction when the prototype is ideal.

6.3. Support Permissibility and Alias Cancellation

From the preceding analysis, we know that each of the three support configurations can satisfy the sampling criterion with two types of $LAT(\mathbf{L}^T)$. Therefore there are six possible choices. If the prototype is an ideal brick-wall filter, perfect reconstruction can be achieved by any of the six choices. When the prototype is not ideal, aliasing will occur in the subbands. The six choices that satisfy the sampling criterion are not all valid candidates in terms of support permissibility [8]. However, in 2D case we have to be more careful in treating the notion of permissibility. We will consider a finer classification of support permissibility: edge-based permissibility and vertex-based permissibility.

Edge-based and vertex-based permissibility. Consider the analysis filter $H_{A,i}(\mathbf{w}) = Q_{A,i}(\mathbf{w}) + Q'_{A,i}(\mathbf{w})$ in configuration A. Suppose an image of $Q_{A,i}(\mathbf{w})$ or $Q'_{A,i}(\mathbf{w})$ is at any of the shaded areas (edge-adjacent to $Q^*_{A,i}(\mathbf{w})$) in Fig. 6.6. If the prototype filter is

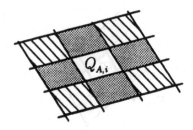

Figure 6.6. Pertaining to the illustration of support permissibility.

ideal, no aliasing is created. However, in practical cases the prototype filter is not ideal. Because of this edge-adjacent image, a considerable amount of edge-based aliasing will be created in the frequency region of the overlapping edges and this edge-based aliasing will add to the major aliasing. Cancellation of this edge-based aliasing is impossible if in the same frequency region there is no other edge-based aliasing error contributing to the same aliasing transfer function [8]. Whenever edge-based aliasing errors do not appear in pairs, the support configuration is called edge-based nonpermissible. Similar to the 1D case, we can argue that the filters in a PR filter bank can not have good stopband attenuation if the support configuration is edge-based nonpermissible. The 2D CMFB studied in [19] usually has this type of nonpermissible support. Similarly, when an image of $Q_{A,i}(\mathbf{w})$ or $Q'_{A,i}(\mathbf{w})$ is at any of the striped areas (vertex-adjacent to $Q^*_{A,i}(\mathbf{w})$) in Fig. 6.6, vertex-based aliasing will occur. If the vertex-based aliasing errors do not appear in pairs, the configuration is called vertex-based nonpermissible. Vertex-based nonpermissibility will also impose limitation on the stopband attenuation of the prototype. But the vertex-based aliasing is, however, not as serious as edge-based aliasing and vertex-based nonpermissibility is not as intolerable as edge-based nonpermissibility.

A simple test of support permissibility. From the general setting described in Sec. 6.1, we know how to obtain the analysis and synthesis filters from a DFT filter bank with twice the number of channels. The supports of the analysis and synthesis filters thus acquired are of a very regular nature. As a result, support permissibility of the two-copy CMFB can be tested through a very simple approach. Consider configuration A in Fig. 6.2. Notice that each analysis filter has two passbands, passband of $Q_{A,i}(\mathbf{w})$ and passband of $Q'_{A,i}(\mathbf{w})$. We can verify that except in special degenerate cases of $\mathbf{N}$, support permissibility of the two-copy CMFB can be described as follows: if $Q_{A,i}(\mathbf{w})$ (or $Q'_{A,i}(\mathbf{w})$) is edge-adjacent to its own images, the configuration is edge-based nonpermissible. Similarly, if $Q_{A,i}(\mathbf{w})$ (or $Q'_{A,i}(\mathbf{w})$) is vertex-adjacent to its own images, the configuration is vertex-based nonpermissible. This testing rule of permissibility applies also to configurations B and C.

Rectangular $LAT(\mathbf{L}^T)$: edge-based nonpermissible support. When $LAT(\mathbf{L}^T)$ is rectangular as in Fig. 6.3(b), we observe that two images of $Q_{A,0}(\mathbf{w})$ are edge-adjacent to $Q_{A,0}(\mathbf{w})$ as shown by shaded areas in Fig. 6.5(a). So configuration A is in general not edge-based permissible with rectangular $LAT(\mathbf{L}^T)$. Similarly, we can show that configuration B

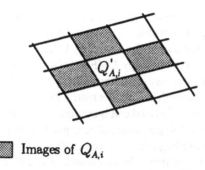

☒ Images of $Q_{A,i}$

Figure 6.7. Images of $Q_{A,i}(\mathbf{w})$ and their positions relative to $Q'_{A,i}(\mathbf{w})$; four images of $Q_{A,i}(\mathbf{w})$ are edge-adjacent to $Q'_{A,i}(\mathbf{w})$.

is not edge-based permissible with rectangular $LAT(\mathbf{L}^T)$ in Fig. 6.3(a). As configuration C can satisfy the sampling criterion only for rectangular $LAT(\mathbf{L}^T)$, configuration C is edge-based nonpermissible and hence not suitable for the development of two-copy CMFB.

Quincunx $LAT(\mathbf{L}^T)$ and edge-based permissibility. *Configuration A and B with quincunx $LAT(\mathbf{L}^T)$.* Consider configuration A first. Location of the images of $Q_{A,0}(\mathbf{w})$ is shown in Fig. 6.5(c). None of the images of $Q_{A,0}(\mathbf{w})$ are edge-adjacent to $Q_{A,0}(\mathbf{w})$. However, images of $Q_{A,0}(\mathbf{w})$ are vertex-adjacent to $Q_{A,0}(\mathbf{w})$. It can be verified that in every subband, the images of $Q_{A,i}(\mathbf{w})$ are vertex-adjacent to $Q_{A,i}(\mathbf{w})$ and images of $Q'_{A,i}(\mathbf{w})$ are vertex-adjacent to $Q'_{A,i}(\mathbf{w})$. In this case, configuration A has edge-based permissibility but lacks vertex-based permissibility.

On the other hand, note that images of $Q_{A,0}(\mathbf{w})$ are edge-adjacent to $Q'_{A,0}(\mathbf{w})$ and result in edge-based aliasing error. Similarly in every other subband, the images of $Q_{A,i}(\mathbf{w})$ are edge-adjacent to $Q'_{A,i}(\mathbf{w})$ and similar edge-based aliasing is created. In particular, it can be shown that four images of $Q_{A,i}(\mathbf{w})$ will be edge-adjacent to $Q'_{A,i}(\mathbf{w})$ as in Fig. 6.7. It turns out that these edge-based aliasing errors are in pairs and can actually cancel with one another if the linear-phase prototype satisfy some minor condition. To be more specific, let the impulse response of the prototype be $p(\mathbf{n})$ and

$$p(\mathbf{n}) = p(\mathbf{n}_s - \mathbf{n}), \text{ for some integer vector } \mathbf{n}_s. \tag{6.2}$$

Cancellation of above-described edge-based aliasing error will take place if

$$\mathbf{n}_s = \mathbf{N}\left(\begin{pmatrix} 0.5 \\ 0.5 \end{pmatrix} + \mathbf{d}\right), \quad \text{for some integer vector } \mathbf{d}. \tag{6.3}$$

This is a minor condition because it is always possible to shift the linear-phase prototype such that (6.3) is satisfied. We conclude that configuration A with quincunx $LAT(\mathbf{L}^T)$ is free from edge-based aliasing if $\mathbf{n}_s$ is as given in (6.3). Applying the preceding analysis on configuration B (Fig. 6.2(b)) we can obtain a similar result: configuration B with quincunx $LAT(\mathbf{L}^T)$ is free from edge-based aliasing if the prototype satisfy the same condition (6.3).

Summary on the construction of the two-copy CMFB. In the above construction, we have set up the framework for two-copy CMFB. Following a very close analogy of 1D CMFB, we proposed the general setting of the two-copy CMFB. In the general setting, we started from a DFT filter bank with decimation matrix $\mathbf{N} = \mathbf{ML}$ for a two-copy CMFB with decimation matrix $\mathbf{M}$. Then we derived configurations A, B and C. For the configurations to satisfy the sampling criterion, the choice of $\mathbf{L}$ is not arbitrary as tabulated in Table 6.1. For those that satisfy the sampling criterion, we examine the support permissibility. In this singling out process, we found that only configuration A and B can possess edge-based permissibility if $LAT(\mathbf{L}^T)$ is quincunx. Configuration C is, however, either violating the sampling criterion or edge-based nonpermissible. Table 6.2 summarizes edge-based permissibility of the two-copy CMFB for each possible combination of $LAT(\mathbf{L}^T)$ and the three configurations. When the two-copy CMFB has edge-based permissibility (configuration A or B with quincunx $LAT(\mathbf{L}^T)$), cancellation of edge-based aliasing is possible. We have seen that edge-based aliasing can be completely cancelled if the prototype has linear phase as in (6.2) and the vector n_s is as in (6.3).

It should be noted that even the best combinations of the configurations and $LAT(\mathbf{L}^T)$ can not achieve both edge-based and vertex-based permissibility. This imposes limitation on the attenuation of the individual filters in the two-copy CMFB as we will see in the design example.

6.4. Perfect Reconstruction Two-Copy CMFB

In this subsection, we present the perfect reconstruction conditions for edge-based permissible two-copy CMFB (configuration A or B with quincunx $LAT(\mathbf{L}^T)$). In the 1D case, the CMFB has perfect reconstruction if and only if the polyphase components of the prototype are pairwise power complementary [26]. We will see now the two-copy CMFB is paraunitary and hence has perfect reconstruction if and only if the polyphase components of the prototype satisfy some 2D power complementary conditions.

The analysis and synthesis filters. From the formulation of the analysis filters $H_m(\mathbf{w})$ in (6.1), the impulse response of $H_m(\mathbf{w})$ assumes the form

$$h_m(\mathbf{n}) = 2p(\mathbf{n})\cos(2\pi(\mathbf{k}_m + \mathbf{b})^T\mathbf{N}^{-1}\mathbf{n}), \mathbf{k} \in \mathcal{N}(\mathbf{N}^T), \quad m = 0, 1, \ldots, J(\mathbf{M}) - 1,$$

$$(6.4)$$

where $\mathbf{b} = (\,0.5 \quad 0\,)^T$ for configuration A and $\mathbf{b} = (\,0 \quad 0.5\,)^T$ for configuration B. As the synthesis filter $F_m(\mathbf{w})$ is given by $F_m(\mathbf{w}) = H_m^*(\mathbf{w})$, the impulse responses of the analysis and synthesis filters are related by

$$f_m(\mathbf{n}) = h_m(-\mathbf{n}), \text{ where } \mathbf{n} \text{ is any } 2 \times 1 \text{ vector.}$$

Let the prototype have the following polyphase representation

$$P(\mathbf{w}) = \sum_{i=0}^{J(\mathbf{N})-1} E_i(\mathbf{N}^T\mathbf{w})e^{-j\mathbf{w}^T\mathbf{n}_i}, \mathbf{n}_i \in \mathcal{N}(\mathbf{N}), \quad (6.5)$$

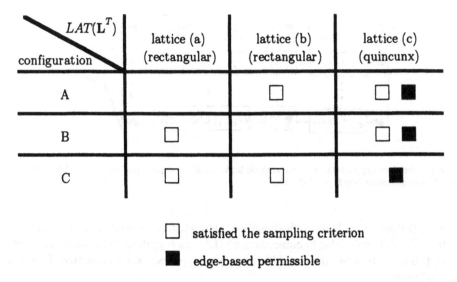

$LAT(\mathbf{L}^T)$ configuration	lattice (a) (rectangular)	lattice (b) (rectangular)	lattice (c) (quincunx)
A		☐	☐ ■
B	☐		☐ ■
C	☐	☐	■

☐ satisfied the sampling criterion

■ edge-based permissible

Table 6.2. The sampling criterion, edge-based permissibility and relation to three possible $LAT(\mathbf{L}^T)$ for configurations A, B and C. Lattice (a)–(c) in the table are as in Fig. 6.3 (a)–(c).

where $E_i(\mathbf{w})$ are the polyphase components of $P(\mathbf{z})$ and the vectors $\mathbf{n}_i \in \mathcal{N}(\mathbf{N})$ are ordered as in (3.2). Then the paraunitariness of the two-copy CMFB can be interpreted in terms of $E_i(\mathbf{w})$.

THEOREM 6.1 *Necessary and sufficient conditions for paraunitariness [31]. Consider the filter bank with decimation matrix* $\mathbf{M}$ *in Fig. 1.1. Let the matrix* $\mathbf{N}$ *be given by* $\mathbf{N} = \mathbf{ML}$, *where* $LAT(\mathbf{L}^T)$ *is quincunx. Choose the analysis filters* $H_m(\mathbf{w})$ *as in (6.4) and let the synthesis filters* $F_m(\mathbf{w}) = H_m^*(\mathbf{w})$. *Also let the prototype be linear phase with* $p(\mathbf{n}) = p(\mathbf{n}_s - \mathbf{n})$ *and* $\mathbf{n}_s = \mathbf{N}(\,0.5 \quad 0.5\,)^T \bmod \mathbf{N}$. *Then the two-copy CMFB is paraunitary (i.e., the polyphase matrix is paraunitary) if and only if*

$$E_i^*(\mathbf{w})E_i(\mathbf{w}) + E_{i+J(\mathbf{M})}^*(\mathbf{w})E_{i+J(\mathbf{M})}(\mathbf{w}) = c, \tag{6.6}$$

where c is some constant.

Remark on Theorem 6.1.   The condition in (6.6) is equivalent to saying that $E_i(\mathbf{w})$ and $E_{i+J(\mathbf{M})}(\mathbf{w})$ are power complementary in 2D sense. This condition can be satisfied by using the 2D paraunitary lattice, [34]. Also the theorem holds for both configuration A and B.

Properties of the two-copy CMFB

1. *The analysis and synthesis filters.* In a 1D CMFB, each analysis filter has two distinct shifts of the prototype filter. So the total bandwidth of each individual filter is the same.

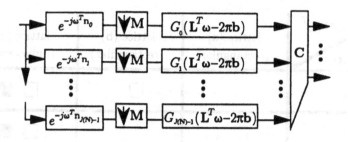

Figure 6.8. Efficient implementation of the analysis bank of the two-copy cosine modulated filter bank. The matrix **C** is of dimension $|\det M|$ by $2|\det M|$.

For 2D filters, total bandwidth should be interpreted as the total spectral occupancy. It can be shown that due to quincunx $LAT(\mathbf{L}^T)$, each analysis filters in the two-copy CMFB consists of two distinct shifts of the prototype and has the same size of spectral occupancy.

2. *Polyphase components of the prototype.* In the 1D CMFB, the polyphase components of the prototype are related in pairs because of linear phase constraint of the prototype. At the same time, there are also power complementary pairs due to paraunitariness. Furthermore if half of the polyphase components are pairwise power complementary, the other half, due to linear phase, are automatically pairwise power complementary, as shown in [26]. The situation is exactly the same in the two-copy CMFB [31].

3. *Efficient implementation of the two-copy CMFB.* Efficient implementation is one of the reasons that cosine modulated filter banks attract a lot of attention. In the 1D CMFB, the complexity of the analysis bank or the synthesis bank is that of the prototype filter plus a DCT matrix. The DCT matrices are known to be low-complexity matrices [56]. There also exists efficient implementation for the two-copy CMFB. The cost of the analysis bank or the synthesis bank is that of a prototype filter plus a matrix, which has elements resembling that of a nonseparable 2D DCT matrix. To be more specific, by using the polyphase representation of the prototype in (6.5), the analysis filters in (6.4) can be rewritten as

$$H_m(\mathbf{w}) = \sum_{i=0}^{J(N)-1} 2E_i(\mathbf{N}^T\mathbf{w} - 2\pi\mathbf{b})[\mathbf{C}]_{mi}\exp(-j\mathbf{w}^T\mathbf{n}_i),$$

$$m = 0, 1, \ldots, J(\mathbf{M}) - 1,$$

where

$$[\mathbf{C}]_{mi} = \cos(2\pi(\mathbf{k}_m + \mathbf{b})^T\mathbf{N}^{-1}\mathbf{n}_i),$$

$$m = 0, 1, \ldots, J(\mathbf{M}) - 1, \quad i = 0, 1, \ldots, J(\mathbf{N}) - 1.$$

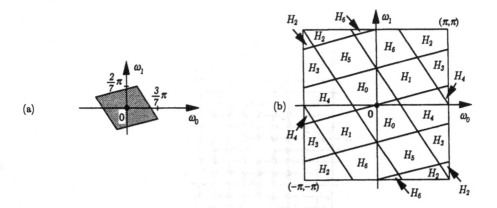

Figure 6.9. Example 6.1. Two-copy cosine modulated filter bank. (a) Spectral support of the prototype $P(\mathbf{w})$, $SPD(\pi \mathbf{N}^{-T})$, and (b) supports of the analysis filters.

The above expression for the analysis filters gives rise to the efficient implementation in Fig. 6.8. The matrix $\mathbf{C}$ is rectangular of dimension $J(\mathbf{M}) \times J(\mathbf{N})$. The figure demonstrates that the complexity of the 2D CMFB is that of the prototype plus $\mathbf{C}$, which can be decomposed into 1D DCT matrices of smaller dimensions.

Example 6.1.   *Two-copy CMFB.* Let $\mathbf{M} = \begin{pmatrix} 7 & -2 \\ 0 & 1 \end{pmatrix}$. Choose $\mathbf{L} = \begin{pmatrix} 1 & 1 \\ 2 & 4 \end{pmatrix}$, then $LAT(\mathbf{L}^T)$ is quincunx. The matrix $\mathbf{N}$ given by $\mathbf{N} = \mathbf{ML}$ is $\mathbf{N} = \begin{pmatrix} 3 & -1 \\ 2 & 4 \end{pmatrix}$. Fig. 6.9(a) shows the support of the prototype. If we choose configuration B, the supports of the analysis filters are as shown in Fig. 6.9(b). By Theorem 6.1, the two-copy CMFB has perfect reconstruction if the polyphase components of the prototype satisfy the power complementary condition given in (6.6). Fig. 6.10 shows the support of impulse response of the prototype filter, $p(n_0, n_1)$. The support of $p(n_0, n_1)$ resembles the shape of $SPD(2\mathbf{N})$. Each solid dot represents a possibly nonzero coefficient of $p(n_0, n_1)$. In this optimization, each of the fourteen polyphase components has four coefficients. The corresponding frequency response of the prototype is shown in Fig. 6.11. The stopband attenuation of the prototype is 17 dB. Since vertex-based permissibility is not achievable in any of our construction, the attenuation of the prototype can not be arbitrary good like in 1D case.

6.5.   Further Discussion on the Construction of the Two-Copy CMFB

In this subsection, we would like to go back to the construction process of two-copy CMFB and have a more detailed discussion of some of the subjects that require a more careful examination.

Support of the prototype. The support of the prototype filter is $SPD(\pi \mathbf{N}^{-T})$. The spectral supports of the individual filters depend on the choice of configurations and the

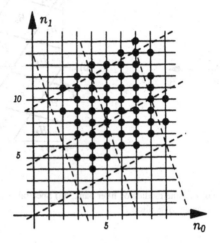

Figure 6.10. Example 6.1. Two-copy cosine modulated filter bank. The impulse response support of the prototype. Each solid dot represents a possibly non-zero coefficient of the prototype. (Intersection points of the dashed lines are on the lattice of **N**. Solid lines represent integers.)

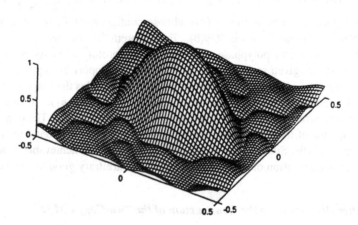

Figure 6.11. Example 6.1. Two-copy cosine modulated filter bank. The magnitude response of the prototype with frequency normalized by 2π.

matrix $\mathbf{N} = \mathbf{ML}$. Although the decimation matrix $\mathbf{M}$ is arbitrary, $\mathbf{L}$ is subject to the constraint that $LAT(\mathbf{L}^T)$ is quincunx. The matrix $\mathbf{N}$ is not arbitrary and hence the support of the prototype is not arbitrary. In the construction of the two-copy CMFB, for a given $\mathbf{M}$ we formulate $\mathbf{N} = \mathbf{ML}$ for some integer matrix $\mathbf{L}$ with quincunx $LAT(\mathbf{L}^T)$. It follows that $\mathbf{N}$ assumes the form

$$\mathbf{N} = \mathbf{M}' \begin{pmatrix} 1 & 1 \\ -1 & 1 \end{pmatrix}, \qquad \text{for some integer matrix } \mathbf{M}'. \tag{6.7}$$

Conversely, it can be shown for any $\mathbf{N}$ of the form in (6.7), we can construct a $J(\mathbf{M}')$ channel two-copy CMFB with decimation matrix $\mathbf{M}'$ such that the prototype filter has support $SPD(\pi\mathbf{N}^{-T})$. However, we should be careful with this statement. Consider

$$\mathbf{N} = \begin{pmatrix} 4 & 0 \\ 4 & 2 \end{pmatrix},$$

which takes the form

$$\mathbf{N} = \underbrace{\begin{pmatrix} 2 & -2 \\ 3 & -1 \end{pmatrix}}_{\mathbf{M}_1} \underbrace{\begin{pmatrix} 1 & 1 \\ -1 & 1 \end{pmatrix}}_{\mathbf{L}_1} \quad \text{or the form} \quad \mathbf{N} = \underbrace{\begin{pmatrix} 4 & 0 \\ 4 & 1 \end{pmatrix}}_{\mathbf{M}_2} \underbrace{\begin{pmatrix} 1 & 0 \\ 0 & 2 \end{pmatrix}}_{\mathbf{L}_2}.$$

Using $SPD(\pi\mathbf{N}^{-T})$ as the prototype support, the 4-channel filter bank with decimation matrix $\mathbf{M}_1$ has edge-based permissibility while the 4-channel filter bank with decimation matrix $\mathbf{M}_2$ does not have edge-based permissibility.

Remark on the choice N=ML. In Sec. 6.2., we have seen that when the prototype $P(\mathbf{w})$ is decimated and then expanded by $\mathbf{M}$, the images are confined to the grid formed by the filters in the DFT filter bank. Furthermore, $P(\mathbf{w})$ and its image form a pattern resembling that of $LAT(\mathbf{L}^T)$. This is a direct result of the choice $\mathbf{N} = \mathbf{ML}$ [31]. If we had chosen $\mathbf{N} = \mathbf{LM}$, then in general the images of $P(\mathbf{w})$ would not be confined to the grid formed by shifts of $P(\mathbf{w})$ in the DFT filter bank.

More general L?. In the preceding construction of two-copy CMFB, we have assumed $\mathbf{L}$ to be an integer matrix of determinant 2. There are only three possible $LAT(\mathbf{L}^T)$. Consider the kth subband in configuration A. From the discussion in Sec. 6.3, we know that $Q_{A,k}(\mathbf{w})$ is always edge-adjacent to its own images for rectangular $LAT(\mathbf{L}^T)$ and $Q_{A,k}(\mathbf{w})$ is always vertex-adjacent to its own images for quincunx $LAT(\mathbf{L}^T)$. In this case the best solution of two-copy CMFB available to us is edge-based permissible but not vertex-based permissible. One wonders whether it is possible to derive a solution of two-copy CMFB that is both edge-based permissible and vertex-based permissible if $\mathbf{L}$ is not restricted to be an integer matrix. The answer is still unfortunately, no! When $\mathbf{L}$ is not an integer matrix but has determinant 2, it can be shown that in some special cases $Q_{A,k}(\mathbf{w})$ is neither edge-adjacent nor vertex-adjacent to any of its own images. In this case, however, the images of passbands of $Q_{A,k}(\mathbf{w})$ will always overlap with passbands of $Q'_{A,k}(\mathbf{w})$; the sampling criterion is violated. The filter bank can not have perfect reconstruction even if all the filters are ideal. This discussion also holds for configuration B and C.

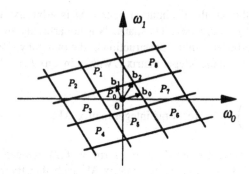

Figure 7.1. DFT filter bank example.

7. Four-Copy Cosine Modulated Filter Banks

In the two-copy CMFB, each analysis filter consists of two copies of the prototype, while in the four-copy CMFB, each analysis filter consists of four copies of the prototype. The general setting of the four-copy CMFB will be an analogy of two-copy CMFB. But some procedures used in the two-copy CMFB require some modifications for the four-copy case (Sec. 7.1). In the two-copy CMFB, we have found that it is not possible to have both edge-based and vertex-based permissibility. In the four-copy case, it is possible but the four-copy CMFB can be edge-based and vertex-based permissible (i.e., completely permissible) only with one particular configuration. This special case of completely permissible four-copy CMFB will be discussed in Sec. 7.2. All the other possible cases of four-copy CMFB will be discussed in Sec. 7.3.

7.1. General Setting

For a four-copy CMFB with decimation matrix $\mathbf{M}$ and $J(\mathbf{M})$ channels, we start from a DFT filter bank with four times the number of channels. Then we shift the filters in the DFT filter bank and combine four shifted filters to obtain real-coefficient analysis filters. However, the procedure of combining four filters to obtain real-coefficient analysis filters is not as straightforward as in the two-copy case.

Similar to the two-copy case, given decimation matrix $\mathbf{M}$, we start from a DFT filter bank with decimation matrix $\mathbf{N} = \mathbf{ML}$, where $\mathbf{L}$ is an integer matrix with $|\det \mathbf{L}| = 4$. Suppose the $4J(\mathbf{M})$-channel DFT filter bank has supports as shown in Fig. 7.1. We can shift the filters in the three directions ($\mathbf{b}_0$, $\mathbf{b}_1$, and $\mathbf{b}_2$) indicated in Fig. 7.1. Fig. 7.2 shows the shifted filters with respect to the three shifts. For instance, if we shift the filters by $\mathbf{b}_2$, the resulting supports are as shown in Fig. 7.2(c). The shifted filters are denoted by $Q_{C,i}$ and $Q'_{C,i}$. Notice that the impulse response of $Q_{C,i}$ and $Q'_{C,i}$ are conjugates of each other. Later when we derive the real-coefficient analysis filters, $Q_{C,i}$ and $Q'_{C,i}$ should always belong to the same analysis filter. The set $(Q_{C,i}, Q'_{C,i})$ will be called a conjugate pair. There is a total of

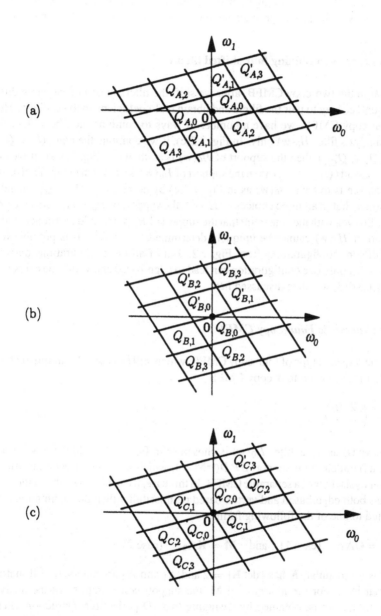

Figure 7.2. Shifted versions of the filters in the DFT filter bank.

$2J(\mathbf{M})$ conjugate pairs. Similarly, in Fig. 7.2(a) and Fig. 7.2(b) $(Q_{A,i}, Q'_{A,i})$ is a conjugate pair and $(Q_{B,i}, Q'_{B,i})$ is a conjugate pair. As each analysis filter in the four-copy CMFB has four copies of the prototype, each analysis filter consists of two conjugate pairs.

The procedure of combining four shifted filters

Recall that in the two-copy CMFB, after we shift the filters in one of the three directions, there is a unique way to pair the filters and obtain real-coefficient analysis filters. However, in the four-copy CMFB, we have a number of ways to combine the filters. Consider the lowpass analysis filter $H_0(\mathbf{w})$ only. In Fig. 7.2(c), if we combine the pair $(Q_{C,0}, Q'_{C,0})$ and the pair $(Q_{C,1}, Q'_{C,1})$, then the support of $H_0(\mathbf{w})$ is as shown in Fig. 7.3(a). If we combine $(Q_{C,0}, Q'_{C,0})$ and $(Q_{C,2}, Q'_{C,2})$, then the support of $H_0(\mathbf{w})$ is as shown in Fig. 7.3(b). Another possible choice is to have $H_0(\mathbf{w})$ as in Fig. 7.3(c) by combining $(Q_{C,0}, Q'_{C,0})$ and $(Q_{C,3}, Q'_{C,3})$. Notice that in all three choices of $H_0(\mathbf{w})$, the support of $H_0(\mathbf{w})$ is connected by edge or vertex. So even with the constraint that the support of $H_0(\mathbf{w})$ should be a connected region, the support of $H_0(\mathbf{w})$ cannot be uniquely determined. Therefore, it is possible to derive several different configurations from Fig. 7.2. But of all the configurations derived from Fig. 7.2(a)–(c), only one configuration can lead to edge-based and vertex-based permissible four-copy CMFB, which is discussed next.

7.2. The Simplistic Four-Copy CMFB

We consider a special type of four-copy CMFB, which will be called the simplistic four-copy CMFB. In the simplistic four-copy CMFB,

$$\mathbf{L} = \begin{pmatrix} 2 & 0 \\ 0 & 2 \end{pmatrix}$$

and the lowpass analysis filter $H_0(\mathbf{w})$ is chosen as in Fig. 7.3(a). So the lowpass analysis filter has a parallelogram support $SPD(\pi\mathbf{M}^{-T})$, which is a natural generalization of the lowpass analysis filter of a separable CMFB. In this case, it can be shown that if the four-copy CMFB has both edge-based and vertex-based permissibility then the decimation matrix $\mathbf{M}$ is restricted to one of the following forms.

$$\mathbf{M} = \mathbf{U}\Lambda_1, \quad \text{(case 1)} \quad \text{and} \quad \mathbf{M} = \mathbf{K}\Lambda_2, \quad \text{(case 2)} \tag{7.1}$$

where $\mathbf{U}$ is unimodular, $\mathbf{K}$ has $|\det \mathbf{K}| = 2$ and Λ_1 and Λ_2 are diagonal. All matrices are integer matrices. For the first case of $\mathbf{M}$, the support configurations of the analysis and synthesis filters can be obtained by designing two 1D perfect reconstruction filter banks and performing a unimodular transformation as explained in Sec. III [49]. If the two 1D filter banks are cosine modulated, then the resulting 2D nonseparable filter bank will also be cosine modulated. That is, all the filters in the 2D nonseparable filter bank can be derived from one prototype. For the second case, the desired support configuration of four-copy CMFB can be achieved by concatenating a separable 2D CMFB with a 2D nonseparable

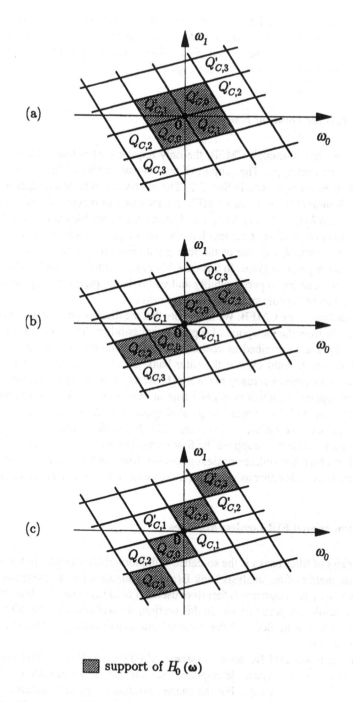

Figure 7.3. Possible supports of $H_0(\mathbf{w})$ when different conjugate pairs are chosen.

two-channel filter bank in the form of a tree structure. We have seen such a tree structure in Example 4.3. In this case, even if the two systems are cosine modulated, the cascaded system is not cosine modulated in general. But the 2D filter banks designed in this method have the desired support configuration of a four-copy CMFB and have perfect reconstruction.

7.3. Other Possible Forms of Four-Copy CMFB

For the most general four-copy CMFB, the only requirement is that each filter contains four copies of the prototype. The support of the lowpass analysis filter is not necessarily a parallelogram as we have seen in Sec. 7.1. The matrix $\mathbf{L}$ can be any integer matrix with $|\det \mathbf{L}| = 4$. Similar to the two-copy CMFB, the locations of images of analysis filters are determined by $LAT(\mathbf{L}^T)$. With $|\det \mathbf{L}^T| = 4$, there are 7 possible choices for $LAT(\mathbf{L}^T)$. Consider the lowpass analysis filter that has spectral support consisting of four connected parallelograms. Namely, any one of the four parallelograms is edge-adjacent or vertex-adjacent to another parallelogram. In this case, it can be verified that such a filter bank can not possess both edge-based permissibility and vertex-based permissibility for any choice of $LAT(\mathbf{L}^T)$, except in some special degenerate cases.

Remarks on four-copy CMFB. We must note that the discussion of alias cancellation is made completely from the viewpoint of support permissibility. Although in the discussion, the analysis filters are described as cosine modulated versions of a prototype filter, the argument continues to hold even if the filter bank is not cosine modulated. In all the discussion, the assumptions actually made are (1) the analysis and synthesis filters have the same spectral support, which is in general true in most filter banks, and (2) the lowpass analysis filter consist of four connected parallelograms. With these two assumptions, we can make the following conclusion. Consider a 2D filter bank, in which the analysis filters contain four parallelograms. Suppose the four parallelograms of the lowpass analysis are connected. If such a filter bank can possess both edge-based permissibility and vertex-based permissibility, then $\mathbf{M}$ is either as given in (7.1) or limited to some degenerate cases.

8.   Two-Dimensional FIR Lossless Matrices

A large subclass of filter banks is the so-called orthonormal filter banks. In these systems, the polyphase matrix of the analysis bank $\mathbf{E}(z)$ is paraunitary, i.e., $\widetilde{\mathbf{E}}(z)\mathbf{E}(z) = \mathbf{I}_M$, [50]. Characterization of such matrices is therefore useful. For 1D systems, this has already been done and the results are well known. In this section, we will consider the 2D counterpart of this topic. Recall from Sec. 1.2 that a causal, stable paraunitary system is also termed as a lossless system.

Characterization of 2D FIR lossless systems. Similar to the 1D case, 2D causal systems can be described by state space description. Several state space realizations have been proposed for 2D causal systems. For the characterization of the 2D lossless systems, the first-level realization [16], [57] and the Roesser realization [40] are of particular importance. The 2D FIR lossless systems have been successfully characterized in terms of first-level

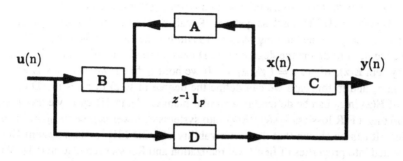

Figure 8.1. The schematic of the state space description for $E(z)$.

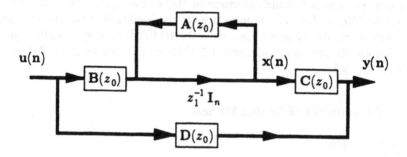

Figure 8.2. The schematic of the first-level description for $E(z)$.

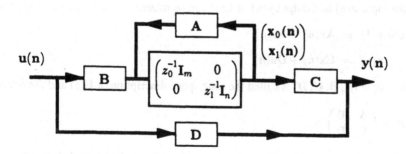

Figure 8.3. The schematic of the Roesser state space description for $E(z)$.

realization [4], [51] and Roesser realization [51]. Losslessness of a 2D system is directly related to the realization matrices in these two types of state space description.

Factorization of 2D FIR lossless systems. Recall in 1D case, any lossless systems can be factorized into degree-one building blocks. To address the issue of factorizing 2D lossless systems, we need to define the degree of a 2D causal transfer matrix first. Consider a 2D causal system $E(z)$, where $z = (z_0 \ z_1)^T$. If we hold z_0 fixed, we can view $E(z)$ as a 1D system in z_1 and in this case we can define the degree of $E(z)$ in z_1 as in 1D case. The degree of $E(z)$ in z_0 can be defined in a similar manner. As in 1D case, we are interested in factorizing a FIR lossless system $E(z)$ into systems of lower degree in z_0, z_1 or in both variables. It can be shown that the factorizability of a 2D FIR lossless system $E(z)$ can be translated into properties of first-level realization and Roesser realization [51]. We will also present a subclass of 2D FIR lossless systems that can be expressed as a product of 1D FIR lossless systems [34].

The main purpose of this section is to review some important developments of 2D FIR lossless systems. A brief review of 1D lossless systems is given in Sec. 8.1. Results on 1D lossless systems will be mentioned only when a related 2D result is to be discussed in this paper. For a more thorough treatment of 1D lossless system, the readers can refer to Chapter 14, [50]. In Sec. 8.2, we introduce the first-level realization [16], [57] and the Roesser realization [40]. Then we characterize the 2D FIR lossless systems through these two realizations. Results on factorization of 2D FIR lossless systems [34], [23], [51], are presented in Sec. 8.3.

8.1. One-Dimensional FIR Lossless Matrices

State space description

Let $E(z)$ be a 1D $M \times M$ causal rational system that can be implemented with p delay elements. Denote the output of the ith delay element by $x_i(n)$ and let

$$\mathbf{x}(n) = (x_0(n) \quad x_1(n) \quad \dots \quad x_{p-1}(n))^T .$$

Then the input $\mathbf{u}(n)$ and output $\mathbf{y}(n)$ of $E(z)$ can be related by the following description.

$$\mathbf{x}(n+1) = \mathbf{A}\mathbf{x}(n) + \mathbf{B}\mathbf{u}(n)$$
$$\mathbf{y}(n) = \mathbf{C}\mathbf{x}(n) + \mathbf{D}\mathbf{u}(n) \tag{8.1}$$

The quadruple $(\mathbf{A}, \mathbf{B}, \mathbf{C}, \mathbf{D})$ is called the state space description of $E(z)$ and the matrix

$$\mathbf{R} = \begin{pmatrix} \mathbf{A} & \mathbf{B} \\ \mathbf{C} & \mathbf{D} \end{pmatrix}$$

is called the realization matrix. It can be shown that any causal rational system $E(z)$ has a state space description as in (8.1). Also from (8.1), we can write $E(z)$ as

$$E(z) = \mathbf{C}(z\mathbf{I} - \mathbf{A})^{-1}\mathbf{B} + \mathbf{D},$$

where $\mathbf{A}$ is of dimension $p \times p$. When p is equal to the degree of $\mathbf{E}(z)$, we say the state space realization is minimal. Fig. 8.1 shows the schematic of the state space description.

In the theorem below, we will see that FIR lossless systems can be characterized in terms of their realization matrices [50].

THEOREM 8.1 *FIR lossless systems and state space realization [46], [14]. A system* $\mathbf{E}(z)$ *is FIR lossless if and only if* $\mathbf{E}(z)$ *admits a minimal realization such that the realization matrix* $\mathbf{R}$ *is unitary, i.e.* $\mathbf{R}^\dagger \mathbf{R} = \mathbf{I}$, *and* $\mathbf{A}$ *is strictly lower triangular. Moreover, the degree of a lossless system* $\mathbf{E}(z)$ *is equal to the degree of* $[\det \mathbf{E}(z)]$.

THEOREM 8.2 *Factorization of FIR lossless matrices [50]. An* $M \times M$ *FIR lossless system* $\mathbf{E}(z)$ *with degree* ρ *can always be factorized as*

$$\mathbf{E}(z) = \mathbf{V}_\rho(z)\mathbf{V}_{\rho-1}(z) \ldots \mathbf{V}_1(z)\mathbf{E}_0, \tag{8.2}$$

where $\mathbf{E}_0$ *is a* $M \times M$ *unitary matrix and* $\mathbf{V}_n(z)$ *is an* $M \times M$ *FIR lossless matrix of degree one given by*

$$\mathbf{V}_n(z) = \mathbf{I} - \mathbf{v}_n\mathbf{v}_n^\dagger + z^{-1}\mathbf{v}_n\mathbf{v}_n^\dagger, \quad \text{with} \quad \mathbf{v}_n^\dagger\mathbf{v}_n = 1. \tag{8.3}$$

8.2. Two-dimensional FIR lossless matrices

8.2.1. First-Level Realization

Let $\mathbf{E}(\mathbf{z})$ be a 2D causal rational system. If we hold z_0 fixed, we can view it as 1D system in z_1, with the coefficients dependent on z_0. Consider a 1D realization of this system and let the state space realization be $(\widehat{\mathbf{A}}(z_0), \widehat{\mathbf{B}}(z_0), \widehat{\mathbf{C}}(z_0), \widehat{\mathbf{D}}(z_0))$. Then the system can be expressed as

$$\mathbf{E}(\mathbf{z}) = \widehat{\mathbf{D}}(z_0) + \widehat{\mathbf{C}}(z_0)(z_1\mathbf{I}_n - \widehat{\mathbf{A}}(z_0))^{-1}\widehat{\mathbf{B}}(z_0). \tag{8.4}$$

This is termed the first-level realization of $\mathbf{E}(\mathbf{z})$. Notice that $\widehat{\mathbf{A}}(z_0)$ is an $n \times n$ matrix function of z_0. The matrix

$$\mathbf{F}(z_0) = \begin{pmatrix} \widehat{\mathbf{A}}(z_0) & \widehat{\mathbf{B}}(z_0) \\ \widehat{\mathbf{C}}(z_0) & \widehat{\mathbf{D}}(z_0) \end{pmatrix}$$

is termed the system matrix associated with this realization. The realization is called minimal if $n = \mu$, where μ is the degree of $\mathbf{E}(\mathbf{z})$ in z_1. The following result is shown in [4], [51].

THEOREM 8.3 *2D FIR lossless systems and first-level realization. The system* $\mathbf{E}(\mathbf{z})$ *with degree* ρ *in* z_0 *and* μ *in* z_1 *is FIR lossless if and only if there exists a minimal first-level realization such that the system matrix* $\mathbf{F}(z_0)$ *is FIR lossless and all the eigenvalues of* $\widehat{\mathbf{A}}(z_0)$ *are 0 for all* z_0. *In this case,* $\mathbf{F}(z_0)$ *has degree* ρ.

8.2.2. Roesser Realization

Roesser proposed the following state space description for 2D causal rational $E(z)$ [5], [40].

$$\begin{pmatrix} x_0(n_0 + 1, n_1) \\ x_1(n_0, n_1 + 1) \end{pmatrix} = A \begin{pmatrix} x_0(n_0, n_1) \\ x_1(n_0, n_1) \end{pmatrix} + Bu(n_0, n_1),$$

$$y(n_0, n_1) = C \begin{pmatrix} x_0(n_0, n_1) \\ x_1(n_0, n_1) \end{pmatrix} + Du(n_0, n_1).$$

So $E(z)$ can be expressed as

$$E(z) = D + C \left(\begin{pmatrix} z_0 I_m & 0 \\ 0 & z_1 I_n \end{pmatrix} - A \right)^{-1} B. \tag{8.5}$$

Similar to 1D case, the realization matrix R is defined as

$$R = \begin{pmatrix} A & B \\ C & D \end{pmatrix}.$$

Let $E(z)$ be of degree ρ in z_0 and μ in z_1. The realization is minimal if $m = \rho$ and $n = \mu$. A schematic of the Roesser realization is shown in Fig. 8.2.

Before we characterize the 2D FIR lossless matrices in terms of the Roesser realization, the definition of the 2D characteristic polynomial is in order. The 2D characteristic polynomial of a $p \times p$ matrix Q with partition $(m, p - m)$ is

$$\alpha_{m, p-m}(z_0, z_1) = \det \left[\begin{pmatrix} z_0 I_m & 0 \\ 0 & z_1 I_{p-m} \end{pmatrix} - Q \right]. \tag{8.6}$$

THEOREM 8.4 [51] *A causal 2D transfer matrix $E(z)$ is FIR lossless if and only if there exists a minimal Roesser realization such that the realization matrix R is unitary and the 2D characteristic polynomial of A satisfies $\alpha_{\rho, \mu}(z_0, z_1) = z_0^\rho z_1^\mu$. Moreover, if $E(z)$ is FIR lossless with degree ρ in z_0 and μ in z_1, then $\det E(z) = c z_0^{-\rho} z_1^{-\mu}$, with $|c| = 1$.*

8.3. Factorization of 2D FIR Lossless Systems

In this subsection, we present some results on the factorization of 2D FIR lossless matrices. We will see that a 2D causal FIR system $E(z)$ can be factorized into FIR lossless systems of lower degree in z_1 if the first-level system matrix is of a certain form (Theorem 8.5). Similarly, we will see that a 2D causal system $E(z)$ can be factorized into FIR lossless systems of lower degree in z_0 and z_1 if the Roesser realization matrix is of a certain form (Theorem 8.6). These two theorems are proved in [51]. Finally, in Theorem 8.7 we will introduce a subclass of 2D FIR lossless systems that can be factorized into the 1D building blocks in (8.3). This subclass consists of 2D FIR lossless systems that have order one in z_0 or z_1 [34].

Before we present the theorems, we look at a structure that leads to 2D FIR lossless systems. A generalization of the factorization in (8.2) to 2D case is [23], [51]

$$\mathbf{E}(\mathbf{z}) = \mathbf{U}_\rho(\mathbf{z})\mathbf{U}_{\rho-1}(\mathbf{z}) \dots \mathbf{U}_1(\mathbf{z})\mathbf{E}_0, \tag{8.7}$$

where $\mathbf{E}_0$ is unitary, and $\mathbf{U}_n(\mathbf{z})$ is a 1D degree-one building block of the form $\mathbf{V}_n(z_0)$ or $\mathbf{V}_n(z_1)$ as defined in (8.3). The systems constructed in this manner are FIR and lossless. However, they only represent a subclass of 2D FIR lossless systems. The reason will follow from Theorem 8.6 below.

THEOREM 8.5 [51] *A 2D FIR lossless system* $\mathbf{E}(\mathbf{z})$ *with degree* μ *in* z_1 *can be factorized as* $\mathbf{E}(\mathbf{z}) = \mathbf{E}_1(\mathbf{z})\mathbf{E}_2(\mathbf{z})$, *where* $\mathbf{E}_i(\mathbf{z})$ *are of degree* μ_i *in* z_1 *and* $\mu = \mu_1 + \mu_2$, *if and only if* $\mathbf{E}(\mathbf{z})$ *admits a minimal first-level realization such that all the following conditions are satisfied.*

1. *The system matrix* $\mathbf{F}(z_0)$ *is FIR and lossless in* z_0.

2. *The matrix* $\widehat{\mathbf{A}}(z_0)$ *in (8.4) is block lower triangular, i.e.*

$$\widehat{\mathbf{A}}(z_0) = \begin{pmatrix} \widehat{\mathbf{A}}_1(z_0) & \underline{0} \\ \times & \widehat{\mathbf{A}}_2(z_0) \end{pmatrix}, \tag{8.8}$$

where $\times$ *represent possibly nonzero polynomial in* z_0^{-1}.

3. *The block matrix* $\widehat{\mathbf{A}}_i(z_0)$ *is of dimension* $\mu_i \times \mu_i$ *and all the eigenvalues of* $\widehat{\mathbf{A}}_i(z_0)$ *are 0 for all* z_0.

A corollary of this theorem is that FIR lossless $\mathbf{E}(\mathbf{z})$ can be factorized into matrices of degree-one in z_1 if and only if $\widehat{\mathbf{A}}(z_0)$ is strictly lower triangular.

THEOREM 8.6 [51] *Let* $\mathbf{E}(\mathbf{z})$ *be a 2D FIR lossless system with degree* ρ *in* z_0 *and* μ *in* z_1. *It can be factorized as* $\mathbf{E}(\mathbf{z}) = \mathbf{E}_1(\mathbf{z})\mathbf{E}_2(\mathbf{z})$, *where* $\mathbf{E}_i(\mathbf{z})$ *are of degree* ρ_i *in* z_0 *and* μ_i *in* z_1 *with* $\rho = \rho_1 + \rho_2$ *and* $\mu = \mu_1 + \mu_2$, *if and only if the following holds:* $\mathbf{E}(\mathbf{z})$ *should admit a minimal Roesser realization* $(\mathbf{A}, \mathbf{B}, \mathbf{C}, \mathbf{D})$ *such that all the following conditions are satisfied.*

1. *The realization matrix* $\mathbf{R}$ *is unitary.*

2. *The matrix* $\mathbf{A}$ *in (8.5) is of the form*

$$\mathbf{A} = \begin{pmatrix} \mathbf{A}_{11} & \underline{0} & \mathbf{A}_{12} & \underline{0} \\ \times & \mathbf{A}_{21} & \times & \mathbf{A}_{22} \\ \mathbf{A}_{13} & \underline{0} & \mathbf{A}_{14} & \underline{0} \\ \times & \mathbf{A}_{23} & \times & \mathbf{A}_{24} \end{pmatrix},$$

where $\times$ *represent possibly nonzero entry.*

3. *The matrix* $\mathbf{A}_i = \begin{pmatrix} \mathbf{A}_{i1} & \mathbf{A}_{i2} \\ \mathbf{A}_{i3} & \mathbf{A}_{i4} \end{pmatrix}$ *with partition* (ρ_i, μ_i) *has characteristic polynomial* $\alpha_{\rho_i, \mu_i}(\mathbf{z}) = z_0^{\rho_i} z_1^{\mu_i}$.

A corollary of this theorem is that FIR lossless $\mathbf{E}(\mathbf{z})$ can be factorized into 1D degree-one building block as in (8.7) if and only if $\mathbf{A}$ is of the form $\mathbf{A} = \mathbf{P}^T \mathbf{L} \mathbf{P}$, where $\mathbf{P}$ is a permutation matrix and $\mathbf{L}$ is strictly lower triangular.

THEOREM 8.7 [34] *Let* $\mathbf{E}(\mathbf{z})$ *be a lossless system of order one in* z_0

$$\mathbf{E}(\mathbf{z}) = \sum_{n=0}^{1} \sum_{n_1=0}^{N} \mathbf{e}(n_0, n_1) z_0^{-n_0} z_1^{-n_1}.$$

Then $\mathbf{E}(\mathbf{z})$ *is of the form* $\mathbf{E}(\mathbf{z}) = \mathbf{G}_0(z_0)\mathbf{F}(z_1)\mathbf{G}_1(z_0)$, *where* $\mathbf{F}(z_1)$, $\mathbf{G}_0(z_0)$ *and* $\mathbf{G}_1(z_0)$ *are 1D FIR lossless systems.*

By Theorem 8.2, $\mathbf{F}(z_1)$, $\mathbf{G}_0(z_0)$ and $\mathbf{G}_1(z_0)$ can be factorized into the 1D building block in (8.3). So the subclass of 2D FIR lossless systems that has order one in z_0 can be expressed as the product of the 1D building block in (8.3). Notice that there is no restriction on the order of z_1.

Remark. Theorem 8.5 and Theorem 8.6 provide the necessary and sufficient conditions for the factorizability of 2D lossless systems. In Theorem 8.5, we see that a 2D lossless system $\mathbf{E}(\mathbf{z})$ can be factorized into systems of lower degrees in z_0 or z_1 if and only if $\mathbf{E}(\mathbf{z})$ admits a minimal first-level realization that satisfies those three conditions described in Theorem 8.5. However, there is no systematic approach to test whether $\mathbf{E}(\mathbf{z})$ admits such a first-level realization. Similarly, Theorem 8.6 states that a 2D lossless system $\mathbf{E}(\mathbf{z})$ can be factorized into systems of lower degrees in z_0 and z_1 if and only if $\mathbf{E}(\mathbf{z})$ admits a minimal Roesser realization that satisfies those three conditions described in Theorem 8.6. Again, there is no systematic approach to test whether $\mathbf{E}(\mathbf{z})$ admits such a Roesser realization.

Acknowledgements

Work supported in parts by NSF grant MIP 92-15785, Tektronix, Inc., and Rockwell International.

References

1. R. Ansari and C. Guillemot, "Exact reconstruction filter banks using diamond FIR filters," *Proc. Int. Conf. on New Trends in Comm. Control, and Signal Proc.*, Turkey, July 1990.

2. R. Ansari and C. L. Lau, "Two-dimensional IIR filters for exact reconstruction in tree-structured subband decomposition," *Electronic Letter*, vol. 23, June 1987, pp. 633–634.

3. R. H. Bamberger and M. J. T. Smith, "A filter bank for the directional decomposition of images: theory and design," *IEEE Trans. on Signal Processing*, vol. SP-40, no. 4, April 1992, pp. 882–893.

4. S. Basu, H. Choi, and C. Chiang, "On non-separable multidimensional perfect reconstruction filter banks," *Proc. of the 27th Annual Asilomar Conference on Signals, Systems and Computers*, 1993, pp. 45–49.

5. N. K. Bose, *Applied Multidimensional Systems Theory*, Van Nostrand Reinhold, 1982.

6. T. Chen and P. P. Vaidyanathan, "The role of integer matrices in multidimensional multirate systems," *IEEE Trans. on Signal Processing*, vol. SP-41, March 1993.

7. T. Chen and P. P. Vaidyanathan, "Recent developments in multidimensional multirate systems," *IEEE Trans. on Circuits And Systems For Video Technology*, vol. 3, no. 2, April 1993, pp. 116–137.

8. T. Chen and P. P. Vaidyanathan, "Consideration in multidimensional filter bank design," *Proc. International Symposium on Circuits and Systems*, May 1993.

9. T. Chen and P. P. Vaidyanathan, "Multidimensional multirate filters and filter banks derived from one dimensional filters," *IEEE Trans. on Signal Processing*, vol. SP-41, May 1993.

10. P. L. Chu, "Quadrature mirror filter design for an arbitrary number of equal bandwidth channels," *IEEE Trans. on Acoustic, Speech and Signal Processing*, vol. 33, Feb. 1985, pp. 203–218.

11. A. Cohen and I. Daubechies, "Nonseparable bidimensional wavelet bases," Preprint, 1993.

12. R. E. Crochiere, S. A. Waber, and J. L. Flanagan, "Digital coding of speech in subbands,"*Bell Sys. Tech. Jour.*, vol. 55, Oct. 1976, pp. 1069–1085.

13. R. E. Crochiere and L. R. Rabiner, *Multirate Digital Signal Processing*, Englewood Cliffs: Prentice Hall, 1983.

14. Z. Doganata, P. P. Vaidyanathan, and T. Q. Nguyen, "General synthesis procedures for FIR lossless transfer matrices, for perfect-reconstruction multirate filter bank applications," *IEEE Trans. on Acoustic, Speech and Signal Processing*, vol. ASSP-36, Oct. 1988, pp. 1561–1574.

15. D. E. Dudgeon and R. M. Mersereau, *Multidimensional Digital Signal Processing*, Englewood Cliffs: Prentice Hall, 1984.

16. R. Eising, "Realization and stabilization of 2-D systems," *IEEE Trans. on Automatic Control*, vol. 23, Oct. 1978, pp. 793–799.

17. B. L. Evans, R. H. Bamberger, and J. H. McClellan, "Rules for multidimensional multirate structures," *IEEE Trans. on Signal Processing*, vol. SP-42, April 1994, pp. 762–771.

18. R. A. Gopinath and C. S. Burrus, "On upsampling, downsampling, and rational sampling rate filter banks," *IEEE Trans. on Signal Processing*, vol. SP-42, April 1994, pp. 812–824.

19. M. Ikehara, "Cosine-modulated 2 dimensional FIR filter banks satisfying perfect reconstruction," *Proc. International Conf. on Acoustic, Speech, and Signal Processing*, vol. III, April 1994, pp. 137–140.

20. M. Ikehara, "Modulated 2 dimensional perfect reconstruction FIR filter banks with permissible passbands," *Proc. International Conf. on Acoustic, Speech, and Signal Processing*, May 1995, pp. 1468–1471.

21. N. S. Jayant and P. Noll, *Digital Coding of Waveforms*, Englewood Cliffs: Prentice Hall, 1984.

22. A. A. C. M. Kalker, "Commutativity of up/down sampling," *Electron. Lett.*, vol. 28, no. 6, March 1992, pp. 567–569.

23. G. Karlsson and M. Vetterli, "Theory of two-dimensional multirate filter banks," *IEEE Trans. on Acoustic, Speech and Signal Processing*, vol. SP-38, June 1990, pp. 925–937.

24. C. W. Kim and R. Ansari, "FIR/IIR exact reconstruction filter banks with applications to subband coding of images," Midwest CAS Symposium, May 1991.

25. C. W. Kim and R. Ansari, "Subband decomposition procedure for quincunx sampling grids," *Proc. SPIE Visual Communications and Image Processing*, Boston, Nov. 1991.

26. R. D. Koilpillai and P. P. Vaidyanathan, "Cosine-modulated FIR filter banks satisfying perfect reconstruction," *IEEE Trans. on Signal Processing*, vol. 40, April 1992, pp. 770–783.

27. J. Kovacevic and M. Vetterli, "The commutativity of up/downsampling in two dimensions," *IEEE Trans. on Information Theory*, vol. 37, no. 4, May 1991, pp. 695–698.

28. J. Kovacevic and M. Vetterli, "Non-separable multidimensional perfect reconstruction filter banks and wavelet bases for R^n," *IEEE Trans. on Information Theory*, vol. 38, no. 2, March 1992, pp. 533–555.

29. J. Kovacevic, "Local cosine bases in two dimensions," *Proc. International Conf. on Acoustic, Speech, and Signal Processing*, vol. IV, May 1995, pp. 2125–2128.

30. M. Kunt, A. Ikonomopoulos and M. Kocher, "Second generation image coding techniques," *Proc. IEEE*, vol. 73, April 1985, pp. 549–574.

31. Y. Lin and P. P. Vaidyanathan, "Theory and design of two-dimensional cosine modulated filter banks," Tech. report, California Institute of Technology, Pasadena, CA, March 1995.

32. Y. Lin and P. P. Vaidyanathan, "Two-dimensional paraunitary cosine modulated perfect reconstruction filter banks," *Proc. International Symposium on Circuits and Systems*, April 1995, pp. 752–755.

33. Y. Lin and P. P. Vaidyanathan, "On the sampling of two-dimensional bandpass signal," In preparation.

34. V. C. Liu and P. P. Vaidyanathan, "On factorization of a subclass of 2D digital FIR lossless matrices for 2D QMF bank applications," *IEEE Trans. on Circuits and Systems*, vol. 37, no. 6, June 1990, pp. 852–854.

35. H. S. Malvar, *Signal Processing with Lapped Transforms*, Norwood, MA: Artech House, 1992.

36. J. H. McClellan, "The design of two-dimensional digital filters by transformations," *Proc. Seventh Annual Princeton Conf. Information Sciences and Systems*, 1973, pp. 247–251.

37. H. J. Nussbaumer, "Pseudo QMF filter bank," *IBM Tech. Disclosure Bulletin*, vol. 24, Nov. 1981, pp. 3081–3087.

38. S. Phoong, C. W. Kim, P. P. Vaidyanathan, and R. Ansari, "A new class of two-channel biorthogonal filter banks and wavelet bases," *IEEE Trans. on Signal Processing*, vol. SP-43, no. 3, March 1995, pp. 649–665.

39. T. A. Ramstad, "Cosine modulated analysis-synthesis filter bank with critical sampling and perfect reconstruction," *Proc. IEEE Int. Conf. Acoustic, Speech and Signal Processing*, Toronto, Canada, May 1991, pp. 1789–1792.

40. R. P. Roesser, "A discrete state-space model for linear image processing," *IEEE Trans. on Automatic Control*, vol. 20, Feb. 1975, pp. 1–10.

41. J. H. Rothweiler, "Polyphase quadrature filters, a new subband coding technique," *Proc. of the IEEE Int. Conf. on Acoustic, Speech and Signal Processing*, April 1973, pp. 1980–1983.

42. I. A. Shah and A. A. C. Kalker, "Generalized theory of multidimensional M-band filter bank design," *EUSIPCO*, 1992, pp. 969–972.

43. I. A. Shah and A. A. C. Kalker, "Algebraic theory of multidimensional filter banks and their design using transformations," preprint.

44. M. J. T. Smith and S. L. Eddins, "Analysis/synthesis techniques for subband image coding," *IEEE Trans. on Acoustic, Speech and Signal Processing*, vol. 38, no. 8, 1990, pp. 1446–1456.

45. D. B. H. Tay and N. G. Kingsbury, "Flexible design of multidimensional perfect reconstruction FIR 2-band filters using transformation of variables," *IEEE Trans. on Image Processing*, vol. 2, no. 4, Oct. 1993, pp. 466–480.

46. P. P. Vaidyanathan, "The discrete-time bounded-real lemma in digital filtering," *IEEE Trans. on Circuits and Systems*, vol. 32, Sept. 1985, pp. 918–924.

47. P. P. Vaidyanathan and T. Q. Nguyen, "A "Trick" for the design of FIR halfband filters," *IEEE Trans. on Circuits and Systems*, vol. 34, March 1987, pp. 297–300.

48. P. P. Vaidyanathan, "Fundamentals of multidimensional multirate digital signal processing," *Sadahana*, vol. 15, Nov. 1990, pp. 157–176.

49. P. P. Vaidyanathan, "New results in multidimensional multirate systems," *Proc. International Symposium on Circuits and Systems*, 1991, pp. 468–471.

50. P. P. Vaidyanathan, *Multirate Systems and Filter Banks*, Englewood Cliffs: Prentice Hall, 1993.

51. S. Venkataraman and B. C. Levy, "State space representations of 2D FIR lossless transfer matrices," *IEEE Trans. on Circuits and Systems*, vol. 41, Feb. 1994, pp. 117–131.

52. M. Vetterli, "Multidimensional subband coding: Some theory and algorithms," *Signal Processing*, vol. 6, no. 2, Feb. 1984, pp. 97–112.

53. E. Viscito and J. P. Allebach, "The analysis and design of multidimensional FIR perfect reconstruction filter banks for arbitrary sampling lattices," *IEEE Trans. on Circuits and Systems*, vol. CAS-38, no. 1, Jan. 1991, pp. 29–41.

54. J. W. Woods and S. D. O'Neil, "Subband coding of images," *IEEE Trans. on Accoust. Speech and Signal Proc.*, vol. 34, Oct. 1986, pp. 1278–1288.

55. J. W. Woods, *Subband Image Coding*, Norwell, MA: Kluwer Academic Publishers, Inc., 1991.

56. P. Yip, and K. R. Rao, "Fast discrete transforms," in *Handbook of Digital Signal Processing*, edited by D. F. Elliott, San Diego, CA: Academic Press, 1987.

57. D. C. Youla, "The synthesis of networks containing lumped and distributed elements," *Proc. Symp. on Generalized Networks*, New York: Polytechnic Institute of Brooklyn Press, Apr. 1966.

Selected References by Topic

Multidimensional Multirate Systems and Subband Coding

[5]   N. K. Bose, 1982.
[52] M. Vetterli, 1984.
[15] D. E. Dudgeon and R. M. Mersereau, 1984.
[54] J. W. Woods and S. D. O'Neil, 1986.
[23] G. Karlsson and M. Vetterli, 1990.
[34] V. C. Liu and P. P. Vaidyanathan, 1990.
[48] P. P. Vaidyanathan, 1990.
[53] E. Viscito and J. P. Allebach, 1991.
[27] J. Kovacevic and M. Vetterli, 1991.
[55] J. W. Woods, 1991.
[28] J. Kovacevic and M. Vetterli, 1992.
[6]   T. Chen and P. P. Vaidyanathan, 1993.
[7]   T. Chen and P. P. Vaidyanathan, 1993.
[4]   S. Basu, H. Choi, and C. Chiang, 1993.
[50] P. P. Vaidyanathan, 1993.
[51] S. Venkataraman and B. C. Levy , 1994.
[17] B. L. Evans, R. H. Bamberger, and J. H. McClellan, 1994.
[18] R. A. Gopinath and C. S. Burrus, 1994.

Two-Dimensional Two-channel Filter Bank Design

[52] M. Vetterli, 1984.
[2]   R. Ansari and C. L. Lau, 1987.

Multidimensional Multirate Systems and Subband Coding

[1] R. Ansari and C. Guillemot, 1990.
[24] C. W. Kim and R. Ansari, 1991.
[45] D. B. H. Tay and N. G. Kingsbury, 1993.
[9] T. Chen and P. P. Vaidyanathan, 1993.
[38] S. Phoong, C. W. Kim, P. P. Vaidyanathan, and R. Ansari, 1995.

Multidimensional Filter Bank Design

[49] P. P. Vaidyanathan, 1991.
[43] I. A. Shah and A. A. C. Kalker, 1992.
[19] M. Ikehara, 1994.
[32] Y. Lin and P. P. Vaidyanathan,1995.
[20] M. Ikehara, 1995.

Multidimensional Systems and Signal Processing, 7, 331–369 (1996)
© 1996 Kluwer Academic Publishers, Boston. Manufactured in The Netherlands.

Local Orthogonal Bases I: Construction

RICCARDO BERNARDINI
EPFL, CH-1015 Lausanne, Switzerland

bernardi@lcavsun1.epfl.ch

JELENA KOVAČEVIĆ
Bell Laboratories, Murray Hill, NJ 07974, USA

jelena@research.bell-labs.com

Abstract. We present a new method for constructing local orthogonal bases, both in continuous and discrete time. All basis functions are obtained from a single prototype window, and there exists a fast algorithm for implementation. The approach is very general and can handle a large variety of cases interesting for applications, such as audio and image coding.

1. Introduction

Discrete-time local cosine bases, or, cosine modulated filter banks[1], have been in use for some time [1], [2]. Due to a few of their properties, they have become quite popular; For example, all filters (basis functions) of a filter bank are obtained by appropriate modulation of a single prototype filter. Then, fast algorithms exist, making them very attractive for implementation. Finally, they have been used recently to achieve time-varying splittings of the time-frequency plane [3]. Their continuous counterpart, termed "Malvar's wavelets", has found use in decomposing a signal into a linear combination of time-frequency atoms [4], [5]. Local cosine bases have been used extensively in audio coding [6]. They have also found use in image coding, due to the reduction of blocking effects [7] when compared to the DCT. Some video works contain local cosine bases as well [8]. However, in all image and video applications, one-dimensional local cosine bases are used separately.

In the multidimensional case one has more freedom than in the one–dimensional one and using only separable cosine bases is very restrictive. In this work we build local orthogonal bases in a very general way by using a formalism that includes the one-dimensional case as well the multidimensional one, the continuous as well the discrete, the separable as well the nonseparable and, moreover, it is able to handle nonrectangular decompositions of the multidimensional domain. The theory presented in this work also allows for a recursive decomposition of the original domain, a tool that could be very useful for applications, since it allows for an adaptative choice of the basis.

2. Review of Local Cosine Bases

Discrete-Time Case: The usual one-dimensional local cosine bases are a class of perfect reconstruction filter banks which use a single prototype filter, window, $w[n]$ of length $2N$

75

(where N is the number of channels and is even) to construct all of the filters $h_0, \ldots, h_{N-1}$ as follows:

$$h_k[n] = \frac{w[n]}{\sqrt{N}} \cdot \cos\left[\frac{2k+1}{4N}(2n - N + 1)\pi\right], \tag{1}$$

with $k = 0, \ldots, N - 1$, $n = 0, \ldots, 2N - 1$, and where the prototype lowpass filter $w[n]$ is symmetric

$$w[n] = w[2N - 1 - n], \qquad n = N, \ldots, 2N - 1 \tag{2}$$

and satisfies [2]

$$w^2[n] + w^2[N - 1 - n] = 2, \qquad n = 0, \ldots, N - 1, \tag{3}$$

which is called the *power-complementarity condition*. This last condition, imposed on the window, ensures that the resulting local cosine basis is orthogonal. The two symmetric halves of the window are called "tails".

Continuous-Time Case: Original construction of local sine and cosine bases in continuous time appeared in [9]. In [10] the same problem is approached in an interesting way. The authors divide the vector space of $V_{\mathbf{R}} = L_2(\mathbb{R})$ into an orthogonal sum of "local" vector spaces.

First, the set $\mathbb{R}$ is decomposed into a union of overlapping intervals $I_i = [\alpha_i - \epsilon_i, \alpha_{i+1} + \epsilon_{i+1}]$ (see Figure 1(a)) such that there is overlapping between I_i and I_{i-1}, for each $i \in \mathbb{Z}$. Note that $I_i \cap I_{i-1} = [\alpha_i - \epsilon_i, \alpha_i + \epsilon_i]$ is centered in α_i. Following the classical point of view, there is a window function w_i associated with each I_i and the perfect reconstruction is guaranteed if the windows enjoy certain properties (such as symmetry and power-complementarity).

Instead, in [10], to each interval I_i is associated a suitable vector space, more precisely, the vector space of the functions obtained via projections of the form

$$\begin{aligned} P_i f(x) = w_i^2(x)f(x) &+ s_{i,-}w_i(x)w_i(2\alpha_i - x)f(2\alpha_i - x) \\ &+ s_{i,+}w_i(x)w_i(2\alpha_{i+1} - x)f(2\alpha_{i+1} - x), \end{aligned} \tag{4}$$

where $w_i(x)$ is a window function with support I_i and $s_{i,-}, s_{i,+} \in \{\pm 1\}$. Observe that $P_i f$ has support I_i. Since P_i is a projection, it is idempotent and self-adjoint (see Appendix B).

The classical constraints of symmetry and power-complementarity on the windows relative to adjacent intervals (in discrete time these are given by (2) and (3)) follow from the conditions of self-adjointness and idempotency of (4) and the condition of orthogonality between two spaces relative to adjacent intervals.

More precisely, the conditions obtained in [10] can be rewritten as

$$\begin{aligned} w_i^2(x) + w_i^2(2\alpha_i - x) + w_i^2(2\alpha_{i+1} - x) &= 1, \quad x \in I_i, \\ \chi_{[\alpha_i - \epsilon_i, \alpha_i + \epsilon_i]} w_i(2\alpha_i - x) &= \chi_{[\alpha_i - \epsilon_i, \alpha_i + \epsilon_i]} w_{i-1}(x), \\ s_{i,-} &= -s_{i-1,+}, \end{aligned} \tag{5}$$

where $\chi[a, b]$ is the indicator function of the interval $[a, b]$, that is, it equals 1 on $[a, b]$ and 0 otherwise. We will now try to give another interpretation to the results from [10] which will be useful for our later developments.

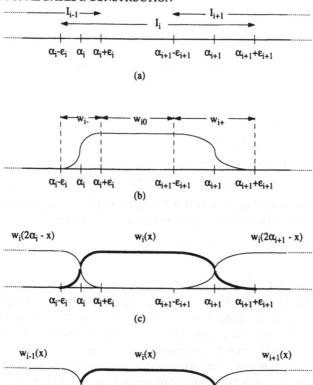

Figure 1. (a) The decomposition of the real line into overlapping intervals I_i [10]. (b) Window $w_i(x)$ on the interval I_i and its pieces $w_{i,-}$, $w_{i,0}$, $w_{i,+}$. (c) The first condition in (5) given pictorially. (d) Windows w_i.

A major insight in conditions (5) can be obtained by writing the window w_i as

$$
\begin{aligned}
w_i(x) &= \chi_{[\alpha_i-\epsilon_i,\alpha_i+\epsilon_i]}w_i(x) + \chi_{[\alpha_i+\epsilon_i,\alpha_{i+1}-\epsilon_{i+1}]}w_i(x) + \chi_{[\alpha_{i+1}-\epsilon_{i+1},\alpha_{i+1}+\epsilon_{i+1}]}w_i(x) \\
&= w_{i,-}(x) + w_{i,0}(x) + w_{i,+}(x),
\end{aligned}
\tag{6}
$$

given in Figure 1(b). In (6) the first and the third term of the sum can be recognized as the left tail and the right tail of the window w_i. By using (6) the first and the second conditions of (5) can be rewritten as (see Figure 1(c) and (d))

$$
\begin{aligned}
w_{i,-}^2(x) + w_{i,-}^2(2\alpha_i - x) &= 1, \\
w_{i,0}^2(x) &= 1, \\
w_{i,+}^2(x) + w_{i,+}^2(2\alpha_{i+1} - x) &= 1, \\
w_{i-1,+}(x) &= w_{i,-}(2\alpha_i - x).
\end{aligned}
\tag{7}
$$

The first three equations of (7) require that the tails of window w_i be power complementary, while the fourth requires that the right tail relative to interval I_{i-1} be obtained from the left tail of window w_i by symmetry around α_i.

The first three equations of (7) can be slightly modified to give them a more intriguing form. Define, for each i, the symmetry $u_i(x) \overset{\Delta}{=} 2\alpha_i - x$ and let $\mathcal{I}(x) = x$. Note that $u_i(u_i(x)) = \mathcal{I}(x)$, that is, $u_i(x)$ is an involution (see Appendix A). Then, (7) can be rewritten as

$$w_{i,-}^2(\mathcal{I}(x)) + w_{i,-}^2(u_i(x)) = 1,$$
$$w_{i,0}^2(\mathcal{I}(x)) = 1,$$
$$w_{i,+}^2(\mathcal{I}(x)) + w_{i,+}^2(u_{i+1}(x)) = 1,$$
$$w_{i-1,+}(x) = w_{i,-}(u_i(x)). \tag{8}$$

A key observation is that, since $u_i(u_i(x)) = \mathcal{I}(x)$, the set $\{\mathcal{I}, u_i\}$ is a group (more precisely a commutative group), the three conditions in (8) can be unified as

$$\sum_{v \in \Gamma[s]} w_{i,s}^2(v(x)) = 1, \qquad s = +, 0, -, \tag{9}$$

where $\Gamma_+ \overset{\Delta}{=} \{\mathcal{I}, u_{i+1}\}$, $\Gamma_- \overset{\Delta}{=} \{\mathcal{I}, u_i\}$ and $\Gamma_0 \overset{\Delta}{=} \{\mathcal{I}\}$.

In (9) each reference to the domain $\mathbb{R}$ disappeared and conditions (5) assume a more abstract and general form. Indeed, as it will become clear later, (9) holds also in the multidimensional case and it can be used both in continuous and discrete time. The main requirement on $\Gamma[s]$ will be that it be a group of involutions.

In [10] a very important characterization of the vector space associated to interval I_i is shown. More precisely, it is proved that the vector space associated with I_i is a set of functions with support I_i. Each of these functions can be written as $f(x) = wi(x)S(x)$, where $S(x)$ is a function, with support I_i, that enjoys the following symmetries:

$$S(x)\chi_{[\alpha_i - \epsilon_i, \alpha_i + \epsilon_i]} = s_{i,-}S(2\alpha_i - x)\chi_{[\alpha_i - \epsilon_i, \alpha_i + \epsilon_i]},$$
$$S(x)\chi_{[\alpha_{i+1} - \epsilon_{i+1}, \alpha_{i+1} + \epsilon_{i+1}]} = s_{i,+}S(2\alpha_{i+1} - x)\chi_{[\alpha_{i+1} - \epsilon_{i+1}, \alpha_{i+1} + \epsilon_{i+1}]}. \tag{10}$$

Note that the above equations mean that the function S is symmetric (antisymmetric) around α_i (or α_{i+1}) on interval $[\alpha_i - \epsilon_i, \alpha_i + \epsilon_i]$ (or $[\alpha_{i+1} - \epsilon_{i+1}, \alpha_{i+1} + \epsilon_{i+1}]$).

Other useful observations can be made by writing for f a decomposition similar to that used for w_i, more precisely

$$\begin{aligned}
f &= f_- + f_0 + f_+ \\
&= \chi_{[\alpha_i - \epsilon_i, \alpha_i + \epsilon_i]} f + \chi_{[\alpha_i - \epsilon_i, \alpha_{i+1} \epsilon_{i+1}]} f + \chi_{[\alpha_{i+1} - \epsilon_{i+1}, \alpha_{i+1} + \epsilon_{i+1}]} f, \\
&= \chi_{[\alpha_i - \epsilon_i, \alpha_i + \epsilon_i]} w_i S + \chi_{[\alpha_i - \epsilon_i, \alpha_{i+1} \epsilon_{i+1}]} w_i S + \chi_{[\alpha_{i+1} - \epsilon_{i+1}, \alpha_{i+1} + \epsilon_{i+1}]} w_i S, \\
&= w_{i,-}\chi_{[\alpha_i - \epsilon_i, \alpha_i + \epsilon_i]} S + w_{i,0}\chi_{[\alpha_i - \epsilon_i, \alpha_{i+1} \epsilon_{i+1}]} S + w_{i,+}\chi_{[\alpha_{i+1} - \epsilon_{i+1}, \alpha_{i+1} + \epsilon_{i+1}]} S, \\
&= w_{i,-}S_- + w_{i,0}S_0 + w_{i,+}S_+.
\end{aligned} \tag{11}$$

We claim that each function f_s, with $s \in \{+, 0, -\}$ is a vector of the space relative to I_i. To prove such an assertion let us verify that, for example, f_- satisfies (10); indeed, $f_- = \chi_{[\alpha_i - \epsilon_i, \alpha_i + \epsilon_i]} w_i S = w_i(x)S_-(x)$ and

$$\begin{aligned}
S_-(x)\chi_{[\alpha_i - \epsilon_i, \alpha_i + \epsilon_i]} &= S(x)\chi_{[\alpha_i - \epsilon_i, \alpha_i + \epsilon_i]}, \\
&= s_{i,-}S(2\alpha_i - x)\chi_{[\alpha_i - \epsilon_i, \alpha_i + \epsilon_i]}, \\
&= s_{i,-}S_-(2\alpha_i - x),
\end{aligned} \tag{12}$$

and the first condition of (10) is verified. Moreover, $\chi_{[\alpha_{i+1}-\epsilon_{i+1},\alpha_{i+1}+\epsilon_{i+1}]}S_- = 0$ and the second condition of (10) is trivially verified.

Such a fact implies that, by calling $\mathcal{V}_{I_i}$ the vector space associated to I_i, $\chi_{[\alpha_i-\epsilon_i,\alpha_i+\epsilon_i]}\mathcal{V}_{I_i}$, $\chi_{[\alpha_i-\epsilon_i,\alpha_{i+1}\epsilon_{i+1}]}\mathcal{V}_{I_i}$ and $\chi_{[\alpha_{i+1}-\epsilon_{i+1},\alpha_{i+1}+\epsilon_{i+1}]}\mathcal{V}_{I_i}$ are subspaces of $\mathcal{V}_{I_i}$ and their sum is $\mathcal{V}_{I_i}$. Moreover, by using a scalar product on $L_2(\mathbb{R})$ (with a generic weight $p(x)$)

$$\langle fg \rangle \triangleq \int p(x)f(x)g^*(x)dx \tag{13}$$

it is easily verified that the three vector spaces $\chi_{[\alpha_i-\epsilon_i,\alpha_i+\epsilon_i]}\mathcal{V}_{I_i}$, $\chi_{[\alpha_i-\epsilon_i,\alpha_{i+1}\epsilon_{i+1}]}\mathcal{V}_{I_i}$ and $\chi_{[\alpha_{i+1}-\epsilon_{i+1},\alpha_{i+1}+\epsilon_{i+1}]}\mathcal{V}_{I_i}$ are mutually orthogonal because their functions have disjoint supports. This implies that

$$\mathcal{V}_{I_i} = \chi_{[\alpha_i-\epsilon_i,\alpha_i+\epsilon_i]}\mathcal{V}_{I_i} \oplus \chi_{[\alpha_i-\epsilon_i,\alpha_{i+1}\epsilon_{i+1}]}\mathcal{V}_{I_i} \oplus \chi_{[\alpha_{i+1}-\epsilon_{i+1},\alpha_{i+1}+\epsilon_{i+1}]}\mathcal{V}_{I_i}, \tag{14}$$

where $\oplus$ denotes the direct sum. Decomposition (14) is interesting, since it can help in proving the orthogonality of two adjacent subspaces and the completeness of their sum.

By multiplying (4) by $\chi_{[\alpha_i-\epsilon_i,\alpha_i+\epsilon_i]}$ one obtains the projection on $\chi_{[\alpha_i-\epsilon_i,\alpha_i+\epsilon_i]}\mathcal{V}_{I_i}$

$$\chi_{[\alpha_i-\epsilon_i,\alpha_i+\epsilon_i]}w_i^2(x)f(x) + s_{i,-}\chi_{[\alpha_i-\epsilon_i,\alpha_i+\epsilon_i]}w_i(x)w_i(2\alpha_i - x)f(2\alpha_i - x). \tag{15}$$

Note that the third term has disappeared since it does not have support on $[\alpha_i - \epsilon_i, \alpha_i + \epsilon_i]$. Equation (15) can be rewritten by using the definition of $u_i(x)$ as

$$\sum_{v \in \Gamma_-} s(v)w_{i,-}(x)w_{i,-}(v(x))f(v(x)) \tag{16}$$

where $s(v)$ is a function from Γ_- onto $\{+1, -1\}$, defined as $s(\mathcal{I}) = 1$ and $s(u_i) = s_{i,-}$. As in the case of (9), we can see that (16) is fairly general and does not use the starting domain $\mathbb{R}$. Indeed, (16) will be used in the following to extend the concept of local orthonormal bases.

In conclusion, in [10] the vector space $L_2(\mathbb{R})$ is written as a direct sum

$$L_2(\mathbb{R}) = \oplus_i \mathcal{V}_{I_i} \tag{17}$$

and each $\mathcal{V}_{I_i}$ is further decomposed according to (14). Figure 2 shows the relationships between several subspaces we mentioned until now.

3. Problem Statement and Solution

In this section we want to mimic the approach in [10] in a more general fashion that fits almost every case of interest. Consider a set X and a vector space $\mathcal{V}_X$ of functions on X and let $\langle \cdot, \cdot \rangle$ be an inner product for $\mathcal{V}_X$. For example, in the one-dimensional continuous-time case $X = \mathbb{R}$ and $\mathcal{V}_X = L_2(\mathbb{R})$; on the other hand, in discrete time $X = \mathbb{Z}$ and $\mathcal{V}_X = \ell_2(\mathbb{Z})$.

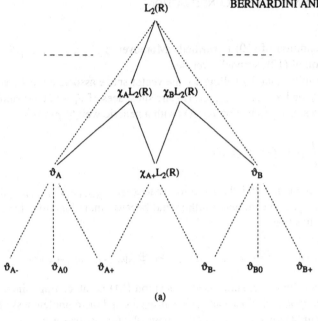

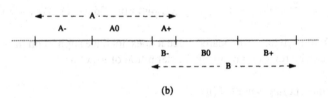

Figure 2. Relationship between subspaces in the decomposition of $L_2(\mathbb{R})$. Solid line denotes that the descendant is a subspace of the parent. All the descendants with dashed lines, joined by a direct sum, form the parent space. For example $V_A = V_{A_-} \oplus V_{A_0} \oplus V_{A_+}$.

In the following we will suppose that

HYPOTHESIS 1

1. *Let f, h, g be functions belonging to V_X, then $\langle f, hg \rangle = \langle f, h^*g \rangle$, as long as the scalar product makes sense.*

2. *Let f be a function of V_X, then $fh \in cV_X$, as long as $|h|$ is a function with finite support and bounded.*

These are two "technical" hypotheses and we need them to justify some derivations, but they are not too restrictive and they are satisfied in most cases of interest. For example, in continuous time the scalar product is often defined as

$$\langle f, g \rangle = \int p(x) f(x) g^*(x) dx, \tag{18}$$

while in discrete time it is of the following form:

$$\langle f, g \rangle = \sum_n p[n] f[n] g^*[n], \tag{19}$$

and thus the first hypothesis is clearly satisfied. As for the second hypothesis, note that it is verified in the two most important cases, that is, $V_X = L_2(\mathbb{R})$ and $V_X = \ell_2(\mathbb{Z})$, because if $|h(x)| \le M_h < \infty$ and $f \in V_X$, then $|f(x)h(x)|^2 \le M_h^2 |f(x)|^2$ and $fh \in V_X$. It is worth noting that the second hypothesis implies that, if C is a bounded subset of X, then $\chi_C V_X$ is a subspace of V_X.

Our goal is to find a local orthonormal basis for V_X, that is, a set of functions $g_{i,k}(x)$ such that

$$f = \sum_{i,k \in \mathbb{Z}} \langle g_{i,k}, f \rangle g_{i,k}, \qquad \forall f \in V_X \tag{20}$$

and each $g_{i,k}$ has finite support B_i. For example, in the one-dimensional discrete-time case,

$$g_{i,k}[n] = h_k[n - Ni] \tag{21}$$

and $B_i = Ni + \{0, \ldots, 2N - 1\}$.

Call V_{B_i} the vector space spanned by $g_{i,k}, k \in \mathbb{Z}$. Vector space V_{B_i}, by construction, has functions with finite support B_i and $g_{i,k}, k \in \mathbb{Z}$ as an orthonormal basis. It is worth noting that V_{B_i} is a subspace of $\chi_{B_i} V_X$, the subspace of V_X of functions with support B_i. In the one-dimensional continuous-time case V_{B_i} is the vector space of the functions $w_i(x) S(x)$, with $S(x)$ satisfying (10). With such an assumption expression (20) is equivalent to

$$V_X = \oplus_i V_{B_i}. \tag{22}$$

It is clear that (22) implies $X = \cup_i B_i$. Therefore, our goal can be achieved as follows:

1. Choose a decomposition $X = \cup_i B_i$.

2. Choose vector spaces V_{B_i} such that (22) is met.

3. Find an orthonormal basis for each V_{B_i}.

The first step is very simple and we will not elaborate on it. In the one-dimensional case one can choose each B_i as an interval, in two dimensions B_i can be, for example, a rectangle. Note that, in general, two sets B_i and B_j will have a nonempty intersection. For example, in the one-dimensional case $B_i \cap B_{i+1} = [\alpha_{i+1} - \epsilon_{i+1}, \alpha_{i+1} + \epsilon_{i+1}]$.

The key point is the second step. To accomplish it we need to investigate the following two points:

1. When are two spaces V_{B_i} and V_{B_j} orthogonal?

2. If they are orthogonal, what is the structure of $V_{B_i} \oplus V_{B_j}$?

The first point is necessary since we want the sum in (22) to be direct and this means that each V_{B_i} is orthogonal to every other V_{B_j}. The second point is motivated by the fact that we want the sum in (22) to be complete or, in other words, we want to be able to write each function of V_X as a sum of functions $h_i \in c V_{B_i}, i \in \mathbb{Z}$.

The answer to these two questions are given in the next two subsections, while the third step (that is, finding an orthonormal basis for each V_{B_i}) is addressed in the last one.

3.1. Orthogonality of V_{B_i} and V_{B_j}

It can be shown that, because of Part 2 of Hypothesis 1, if $B_i \cap B_j = \emptyset$ then $V_{B_i} \perp V_{B_j}$, so we need to study only the case when $B_i \cap B_j = C \neq \emptyset$. In the one-dimensional continuous-time case, we have to study only the case relative to two adjacent intervals $[\alpha_0 - \epsilon_0, \alpha_1 + \epsilon_1]$ and $[\alpha_{-1} - \epsilon_{-1}, \alpha_0 + \epsilon_0]$, where the set C is $[\alpha_0 - \epsilon_0, \alpha_0 + \epsilon_0]$.

Suppose that $\chi_C V_{B_i}$ is a subspace of V_{B_i} (this requirement will be satisfied by construction) and define $D_i = B_i \setminus C$. It is easy to see that, because of Hypothesis 1, $\chi_{D_i} V_{B_i}$ and $\chi_C V_{B_i}$ are orthogonal. Indeed, if $f \in \chi_C V_{B_i}$ and $g \in \chi_{D_i} V_{B_i}$ we have that $f = \chi_C f$ and $g = \chi_{D_i} g$ and $\langle f, g \rangle = \chi_C f \chi_{D_i} g$ that can be rewritten, because of Hypothesis 1, as $\langle \chi_C \chi_{D_i} f, g \rangle$. Since $C \cap D_i = \emptyset$, then $\chi_C \chi_{D_i} = 0$ from which the orthogonality of f and g follows.

Moreover, since $\chi_{B_i} = \chi_C + \chi_{D_i}$, each function $f \in V_{B_i}$ can be written as $f = \chi_{B_i} f = \chi_C f + \chi_{D_i} f$, that is, the sum of a function from $\chi_C V_{B_i}$ with a function from $\chi_{D_i} V_{B_i}$. Such a fact, together with the orthogonality property and the hypothesis that $\chi_C V_{B_i}$ is a subspace of V_{B_i}, implies that $V_{B_i} = \chi_C V_{B_i} \oplus \chi_{D_i} V_{B_i}$. A completely analogue reasoning holds for V_{B_j}. It is clear that the question about orthogonality of V_{B_i} and V_{B_j} becomes about orthogonality of $\chi_C V_{B_i}$ and $\chi_C V_{B_j}$.

In the one-dimensional case $\chi_{[\alpha_0 - \epsilon_0, \alpha_0 + \epsilon_0]} V_{[\alpha_0 - \epsilon_0, \alpha_1 + \epsilon_1]}$ is the vector space of functions $f = wS$ with S having support $C = [\alpha_0 - \epsilon_0, \alpha_0 + \epsilon_0]$ and symmetric or antisymmetric with respect to α_0, depending on the sign chosen in (10). Moreover, w restricted to $[\alpha_0 - \epsilon_0, \alpha_0 + \epsilon_0]$ enjoys the property

$$w^2(x) + w^2(2\alpha - x) = 1, \tag{23}$$

for each x belonging to $[\alpha_0 - \epsilon_0, \alpha_0 + \epsilon_0]$. In a more general case we can have more than one symmetry. For example, in the two-dimensional case we can have the symmetry u_x with respect to the x axis, the symmetry u_y with respect to the y axis and they can be combined to obtain u_{xy} that is, the symmetry with respect to the origin. Note that $\Gamma = \{\mathcal{I}, u_x, u_y, u_{xy}\}$, where $\mathcal{I}(x) = x$, is a group and each element is an involution. To give the type of the symmetry (if even or odd) one can assign to each symmetry u a sign $s(u) = \pm 1$. It can be readily seen that, in order to have coherence between the types of symmetries, if $u, v \in \Gamma$ then it must be $s(uv) = s(u)s(v)$, i.e., s must be a representation of Γ. Some properties of involution groups and their representations are given in Appendix A. Note that in the one-dimensional case Γ has only two elements. With such a point of view, the property (23) can be rewritten as $\sum_{u \in \Gamma} w^2(u(x)) = 1$ as we have already seen in (9), where Γ is a finitely generated abelian group of involutions.

Inspired by these observations we can impose the following structure on the vector space $\chi_C V_{B_i}$:

DEFINITION 1 *Let us make the following assumptions:*

- *Let V_C be the subspace of V_X formed by the functions of V_X with support on C.*

- *Let Γ be a group of involutions on C. Each involution induces a transformation on functions of V_C as $f_u \overset{\Delta}{=} f(u(x))$.*

- *Suppose that $f \in V_C \Rightarrow fu \in V_C$ and that $\langle a, b \rangle = \langle a_u, b_u \rangle$.*

- *Let s be a function mapping Γ onto $\{+1, -1\}$ such that $s(uv) = s(u)s(v), \forall u, v \in \Gamma$, i.e., s is a representation of order two of Γ.*

- *Finally, suppose that w is a function with support C such that*

$$\sum_{u \in \Gamma} w^2(u(x)) = 1. \tag{24}$$

Then we define the vector space V_C as the space of the functions f of V_C such that f can be written as wS, where S is a function with support C satisfying

$$S_u(x) = S(u(x)) = s(u)S, \tag{25}$$

that is, S will possess certain symmetry properties.

It is possible to prove that the orthogonal projection from V_X to $V_C(\Gamma, s, w)$ is

$$P(\Gamma, s, w)f \overset{\triangle}{=} \sum_{u \in \Gamma} s(u)ww_u f_u, \tag{26}$$

and the proof is given in Appendix C.

It is worth noting that the problems of the orthogonality and completeness can be rephrased in term of the corresponding projections, as we will do.

If we impose on $\chi_C V_{B_i}$ the structure described by Definition 1 (we can, since we are designing V_{B_i}), the solution to the problem of orthogonality is included in the following property:

PROPERTY 1 *Let $P_1 \overset{\triangle}{=} P(\Gamma, s, w)$ and $P_2 \overset{\triangle}{=} P(\Gamma, t, w_v)$ where $v \in \Gamma$, $t(v) = -s(v)$. Then $P_1 \perp P_2$.*

The proof of this property is given in Appendix C.

In the one-dimensional case, for $C = [\alpha_0 - \epsilon_0, \alpha_0 + \epsilon_0]$, there is only one nontrivial symmetry (that is, $u_0(x) = 2\alpha_0 - x$) and Property 1 asserts that the right tail of the window of $[\alpha_{-1} - \epsilon_{-1}, \alpha_{-1+1} + \epsilon_{-1+1}]$ must be obtained from the left one of $[\alpha_0 - \epsilon_0, \alpha_1 + \epsilon_1]$ with a symmetry around α_0 and that the signs of the two projections must be different, as given in Proposition 1 of [10] and given, respectively, in the last equation of (8) and in the last equation of (5). However, Property 1 holds also in the multidimensional case.

3.2. Structure of $V_{B_i} + V_{B_j}$

In order to solve the problem of completeness suppose that a given set C is contained in sets $B_1, B_2, \ldots, B_M$ and it has empty intersection with the others. For the relationship (22) to hold it must be

$$\chi_C V_X = \oplus_{i=1}^M \chi_C V_{B_i} \tag{27}$$

or, in other words, by summing functions in $\chi_C V_{B_i}, i = 1, \ldots, M$ one must be able to obtain every function of V_X with support C. If $\chi_C V_{B_i}$ has the structure described by Definition 1, the following property gives a solution to the problem of completeness in the specific case when there is only one symmetry beside the trivial one:

PROPERTY 2 *Suppose* $\Gamma = \{\mathcal{I}, v\}$ *and let* $P_1 \overset{\Delta}{=} P(\Gamma, s, w)$ *and* $P_2 \overset{\Delta}{=} P(\Gamma, t, w_v)$ *where* $t(v) = -s(v)$. *Then* $P_1 \perp P_2$ *and* $P_1 f + P_2 f = f, \forall f \in \chi_C V_X$.

First part of Property 2 is a particular case of Property 1 and ensures the orthogonality of $V_1 = V_C(\Gamma, s, w)$ and $V_2 = V_C(\Gamma, t, w_v)$, that is, the sum of V_1 and V_2 is direct. The second part claims that in this particular case, where Γ has only one nontrivial symmetry, the sum $V_1 \oplus V_2$ is also complete, that is, $V_1 \oplus V_2 = \chi_C V_X$ or, in other words, each function of V_X having support C can be expressed as the sum of a function of V_1 and a function of V_2.

Since in the one-dimensional case each transition zone $[\alpha_0 - \epsilon_0, \alpha_0 + \epsilon_0]$ has only one symmetry, Property 2 is sufficient to study the orthogonality and completeness of a given decomposition in the one-dimensional case.

In the spirit of Property 2 let us give the following definition:

DEFINITION 2 *Two projections satisfying the hypothesis of Property 2 will be called com-plementary to one another. The complementary projection to a projection P will be denoted as P^* (not to be confused with the adjoint).*

In order to handle the case in which Γ has more than one symmetry, we need a tool to "split" a projection as a chain of simpler projections; such a tool is furnished by the two properties shown below; they can be seen as inverses of each other. These properties have no counterpart in the one-dimensional case and to give an example we must resort to multiple dimensions. The first property says, in a formal way, an intuitively appealing fact: if we perform, in sequence, first a projection with a certain symmetry (for example, even symmetry around x axis) then another with another symmetry (for example, odd symmetry around y axis) we obtain a signal with a combination of the two symmetries; in this case we would expect a signal that is even symmetric with respect to the x axis, odd symmetric with respect to the y axis and odd symmetric with respect to the origin, since this symmetry is the combination of an odd symmetry and an even one.

PROPERTY 3 *Let* $P_1 \overset{\Delta}{=} P(\Gamma_1, s, w)$ *and* $P_2 \overset{\Delta}{=} P(\Gamma_2, t, y)$. *Suppose that* $\Gamma_1 \cap \Gamma_2 = \mathcal{I}$, *that* $< \Gamma_1, \Gamma_2 >$ *is a group of involutions and that for each* $v \in \Gamma_2$ *and* $u \in \Gamma_1$ *the following holds*

$$w_v = w, \qquad y_u = y. \tag{28}$$

Then, operator $P_1 P_2$ *is a projection and, moreover*

$$P_1 P_2 = P(< \Gamma_1, \Gamma_2 >, s \otimes t, wy) \tag{29}$$

The proof of this property is given in Appendix C.

Finally, the last property is the inverse of Property 3 and it tells us when a given projection can be expressed as a composition of two simpler projections. It can be easily seen that (29) imposes some structure on the window of the resulting projections. It is clear that in order to invert Property 3 we need some hypothesis on the window.

PROPERTY 4 *Let $P(\Gamma, s, w)$ be a projection and suppose that Γ_1 and Γ_2 are two subgroups of Γ such that $\Gamma =< \Gamma_1, \Gamma_2 >$. Also let r and t be two representations of Γ_1 and Γ_2, respectively, such that $s = r \otimes t$. Suppose that window w is such that $w w_{uv} = w_u w_v$, for each $u \in \Gamma_1$, $v \in \Gamma_2$.*

Then, projection $P(\Gamma, s, w)$ can be written as

$$P(\Gamma, s, w) = P(\Gamma_1, r, \theta) P(\Gamma_2, t, r) \qquad (30)$$

where θ and τ are defined as

$$\theta = \frac{w}{\sqrt{\sum_{u \in \Gamma_1} w_u{}^2}}, \qquad \tau = \frac{w}{\sqrt{\sum_{v \in \Gamma_2} w_v{}^2}}. \qquad (31)$$

In order to complete the construction of basis (20), one has to find an orthonormal basis for $\mathcal{V}_{B_i}$. Note that $\mathcal{V}_{B_i}$ can be written as the direct sum of its subspaces $\mathcal{V}_{C_{i,j}}(\Gamma, s, w)$ relative to the subsets $C_{i,j}$ forming B_i. Using such a fact, an orthonormal basis for $\mathcal{V}_{B_i}$ can be found by finding an orthonormal basis $b_{i,j,k}$, $k \in \mathbb{N}$, for each $\mathcal{V}_{C_{i,j}}$; an orthonormal basis for B_i will be $\{b_{i,1,k_1}, b_{i,2,k_2}, \ldots\}$ with $k_i \in \mathbb{Z}$. Thus, we can limit ourselves to the study of orthonormal bases of $\mathcal{V}_{C_{i,j}}$, as it is done in the next section. For convenience we will call $C_{ij} = C$ and $\mathcal{V}_{C_{i,j}}(\Gamma, s, w) = \mathcal{V}_C(\Gamma, s, w)$.

3.3. An Orthonormal Basis: Discrete Case

Finally, we address Point 3 in Section 3. The problem of finding an orthonormal basis for $\mathcal{V}_C(\Gamma, s, w)$ is simplified by the property presented below. The intuitive meaning is that $\mathcal{V}_C(\Gamma, s, w)$ is constructed by multiplying functions having certain symmetries by the window w and, for this reason, one can expect that a basis for $\mathcal{V}_C(\Gamma, s, w)$ could be obtained by multiplying by w a basis of the space of functions opportunely symmetric on C. Note that such a space can not be written as $\mathcal{V}_C(\Gamma, s, \chi_C)$ since χ_C does not satisfy the condition of normalization of the power; instead one has to write $\mathcal{V}_C(\Gamma, s, \chi_C/\sqrt{|\Gamma|})$, that is, with a scaled window which we will call $w_C = \chi_C/\sqrt{|\Gamma|}$, where $|\Gamma|$ is the cardinality of Γ, that is, the number of elements of Γ.

PROPERTY 5 *Let b_i be an orthonormal basis of $\mathcal{V}(\Gamma, s, w_C)$. Then $\sqrt{|\Gamma|} w b_i$ is an orthonormal basis of $\mathcal{V}_C(\Gamma, s, w)$.*

The proof of this property is given in Appendix C.

For simplicity, here we discuss the problem of finding an orthonormal basis for $\mathcal{V}_C(\Gamma, s, w_C)$ only in the case of C discrete and finite, i.e., when $\mathcal{V}_C$ has a finite dimension. Similar reasoning could be used in the continuous-time case, and will be presented in Section 3.4.

What we need to find out now, is how many free parameters we have in the system. Thus, we will first examine the dependencies that exist. Let us digress for a moment and define two concepts we will use in the remainder of the paper.

DEFINITION 3 *Let C be a set and let Γ be a group of functions mapping C in itself. To each point $x \in C$ one can associate the set of points obtained by applying on x the symmetries of Γ, that is, the set*

$$O(x) \stackrel{\Delta}{=} \{y : y = u(x), u \in \Gamma\}. \tag{32}$$

Set $O(x)$ is called the orbit *of x.*

Note that $x \in O(x)$, since $\mathcal{I} \in \Gamma$. It is possible to show [11] that the orbits relative to the points of C constitute a partition of C.

DEFINITION 4 *Let C be a set and let Γ be a group of functions mapping C in itself. Consider $x \in C$. The set of functions of Γ that map x in itself*

$$Stab(x) \stackrel{\Delta}{=} \{u \in \Gamma : u(x) = x\} \tag{33}$$

is called the stabilizer *of x.*

It is easy to show [11] that $Stab(x)$ is a subgroup of Γ.

If $f \in \mathcal{V}_C(\Gamma, s, w_C)$, then, by definition, $f(u(x)) = s(u)f(x)$, that is, the values assumed by f on a given orbit $O(x)$ are not independent, rather they are determined by the value assumed by f on any element of $O(x)$. It is worth noting that if for some $x \in C$ there is an $u \in \Gamma$ such that $u(x) = x$ and $s(u) = -1$, then $f(x) = f(u(x)) = s(u)f(x) = -f(x) = 0$, that is, f is zero on each point of $O(x)$.

By partitioning C in orbits and choosing one representative from each orbit, a set of representatives $\tilde{C}$, can be made. Note that each function belonging to $\mathcal{V}(\Gamma, s, w_C)$ is uniquely determined by the values assumed on $\tilde{C}$.

For example, in the one-dimensional case it is worth noting that a function of $\mathcal{V}_C(\Gamma, s, w)$, where C is the transition zone $[\alpha_0 - \epsilon_0, \alpha_0 + \epsilon_0]$, is uniquely determined by its values on $\tilde{C} = [\alpha_0, \alpha_0 + \epsilon_0]$, since the other values can be obtained by symmetry. Since an analog reasoning holds also for $[\alpha_1 - \epsilon_1, \alpha_1 + \epsilon_1]$ it is easy to see that a function belonging to $\mathcal{V}_{[\alpha_0 - \epsilon_0, \alpha_1 + \epsilon_1]}$ is determined by its values on $[\alpha_0 - \epsilon_0, \alpha_1 + \epsilon_1] = [\alpha_0, \alpha_1]$.

By discarding all x on which f is identically zero (we can do so, since such points do not add information to f and we can "restore" them whenever we wish), we can form a vector by ordering the points of C "by orbits", i.e., by putting first the points of one orbit, then the points of another orbit, and so on. Only one value of those belonging to a "subvector" corresponding to a particular orbit is free. We will choose as the free parameter the first one, the others will be obtained by multiplying it by ± 1, according to the sign of $x(u)$.

As an example, suppose that

- C is the square $[-2 \ldots 2] \times [-2 \ldots 2]$;

Figure 3. Orbits of square $[-2\ldots2] \times [-2 \times 2]$. Same symbols denote points belonging to the same orbit. The group of symmetries is $\Gamma = \{\mathcal{I}, u_x, u_y, u_{xy}\}$ taken around the central point.

- group Γ is $\{\mathcal{I}, u_x, u_y, u_{xy}\}$, where u_x, u_y and u_{xy} are the symmetries around the horizontal axis, the vertical axis and the origin;

- representation s is given by $s(\mathcal{I}) = s(u_x) = 1$ and $s(u_y) = s(u_{xy}) = -1$.

Set C is depicted in Figure 3. As it is easily seen, the orbits in which C is partitioned are (the origin $(0, 0)$ is in the central point):

$$
\begin{aligned}
O_0 &= O[0, 0] &&= \{[0, 0]\}, \\
O_1 &= O[0, 1] &&= \{[0, 1], [0, -1]\}, \\
O_2 &= O[0, 2] &&= \{[0, 2], [0, -2]\}, \\
O_3 &= O[1, 0] &&= \{[1, 0], [-1, 0]\}, \\
O_4 &= O[2, 0] &&= \{[2, 0], [-2, 0]\}, \\
O_5 &= O[1, 1] &&= \{[1, 1], [-1, 1], [1, -1], [-1, -1]\}, \\
O_6 &= O[1, 2] &&= \{[1, 2], [-1, 2], [1, -2], [-1, -2]\}, \\
O_7 &= O[2, 1] &&= \{[2, 1], [-2, 1], [2, -1], [-2, -1]\}, \\
O_8 &= O[2, 2] &&= \{[2, 2], [-2, 2], [2, -2], [-2, -2]\}.
\end{aligned}
\tag{34}
$$

Note that the functions of $\mathcal{V}_C(\Gamma, s, w)$ are zero on the vertical axis because $s(u_y) = -1$ and the points of vertical axis map in themselves with the function u_y. Therefore, we will exclude from our reasoning the first three orbits of (34). In this case, $\mathcal{V}_C(\Gamma, s, w)$ can be seen as a subspace of dimension 6 (the number of orbits on which the functions of $\mathcal{V}_C(\Gamma, s, w)$ are not identically zero) of $\mathbf{R}^{|C|} = \mathbf{R}^{25}$. By resorting to the ordering suggested before, a vector f of $\mathcal{V}_C(\Gamma, s, w)$ can be written as

$$
f = \mathbf{W} \begin{bmatrix} f_3 \\ f_4 \\ f_5 \\ f_6 \\ f_7 \\ f_8 \end{bmatrix}
\tag{35}
$$

where block f_i contains the values of f on orbit O_i and $\mathbf{W}$ is a diagonal matrix having on the main diagonal the values of the window. In order to organize the values in a single block f_i one can choose an ordering of elements of Γ, for example $\mathcal{I}, u_x, u_y, u_{xy}$, and write the components of f_i according to such an ordering, that is,

$$f_i = \begin{bmatrix} f(p_i) \\ f(u_x(p_i)) \\ f(u_y(p_i)) \\ f(u_{xy}(p_i)) \end{bmatrix}. \tag{36}$$

where p_i is the chosen representative of O_i. However, note that orbits O_3 and O_4 have only two points. This is because their representatives ($[1, 0]$ and $[2, 0]$) have a nontrivial stabilizer; more precisely, $\text{Stab}([1, 0]) = \text{Stab}([2, 0]) = \{\mathcal{I}, u_x\}$. For this reason blocks f_3 and f_4 should have two components, that is

$$f_3 = \begin{bmatrix} f([1, 0]) \\ f([-1, 0]) \end{bmatrix} \qquad f_4 = \begin{bmatrix} f([2, 0]) \\ f([-2, 0]) \end{bmatrix}. \tag{37}$$

However, it is convenient to store values in f_3 and f_4 according to (36), that is,

$$f_3 = \begin{bmatrix} f([1, 0]) \\ f([1, 0]) \\ f([-1, 0]) \\ f([-1, 0]) \end{bmatrix} \qquad f_4 = \begin{bmatrix} f([2, 0]) \\ f([2, 0]) \\ f([-2, 0]) \\ f([-2, 0]) \end{bmatrix}. \tag{38}$$

This does not change the structure of our vector space and simplifies the theoretical reasoning.

Expression (36) now becomes

$$f_i = \begin{bmatrix} f(p_i) \\ s(u_x)f(p_i) \\ s(u_y)f(p_i) \\ s(u_{xy})f(p_i) \end{bmatrix} = \begin{bmatrix} f(p_i) \\ f(p_i) \\ -f(p_i) \\ -f(p_i) \end{bmatrix} = f(p_i) \begin{bmatrix} 1 \\ 1 \\ -1 \\ -1 \end{bmatrix} = f(p_i)K. \tag{39}$$

Substituting (39) into (35) one obtains a general structure for a vector of $\mathcal{V}_C$

$$f = \mathbf{W}(I_6 \otimes K) \begin{bmatrix} f([1, 0]) \\ f([2, 0]) \\ f([1, 1]) \\ f([1, 2]) \\ f([2, 1]) \\ f([2, 2]) \end{bmatrix} = \mathbf{W} \, \text{diag}\{K, K, K, K, K, K\} \begin{bmatrix} f([1, 0]) \\ f([2, 0]) \\ f([1, 1]) \\ f([1, 2]) \\ f([2, 1]) \\ f([2, 2]) \end{bmatrix} \tag{40}$$

where I_6 is the 6×6 identity matrix and $\otimes$ denotes the Kronecker product. Note that the column vector in (40) has no constraints.

This reasoning can be repeated in the general case and it is easy to see that each $f \in \mathcal{V}_C$ can be written as

$$f = \mathbf{WK}g = \mathbf{W}\left(\mathbf{I}_{N_o} \otimes K\right)g \tag{41}$$

where N_O is the number of orbits of Γ on which f is not zero, g is a vector without constraints and $\mathbf{K}$ is the block diagonal matrix $\mathrm{diag}\{K, K, \ldots, K\}$, where K is a column vectors with entries ± 1 and $\mathbf{W}$ contains the values of the window suitably ordered. It is easy to see that $(\mathbf{WK})^T \mathbf{WK} = \mathbf{I}$ as long as the window satisfies the power-complementarity conditions.

To verify such a fact partition $\mathbf{W}$ as $\mathbf{W} = \mathrm{diag}\, W_1, W_2, \ldots, W_M$ where W_i is a diagonal matrix corresponding to ith K on $\mathbf{I}_{|\Gamma|} \otimes K$. With such a notation the product $\mathbf{K}^T \mathbf{WWK}$ can be expressed as $\mathrm{diag}\, K^T W_1^2 K, \ldots K^T W_M^2 K$. By calling $u_1 = \mathcal{I}, u_2, \ldots u_{|\Gamma|}$ the elements of Γ, the matrix product $K^T W_i K$ can be written as

$$\begin{bmatrix} s(u_1) & \cdots & s(u_{|\Gamma|}) \end{bmatrix} \begin{bmatrix} w^2[x] & & \\ & \ddots & \\ & & w^2[u_{|\Gamma|}[x]] \end{bmatrix} \begin{bmatrix} s(u_1) \\ \vdots \\ s(u_{|\Gamma|}) \end{bmatrix}$$

$$= \sum_{i=0}^{|\Gamma|} s(u_i)^2 w^2[u_i[x]] \tag{42}$$

and the last sum is 1 according to the power-complementarity condition.

Suppose now that vectors $c_i, i = 1, \ldots, M$ are an orthonormal basis of $\mathcal{V}_C$. Let V be the matrix having as columns the vectors c_i. Observe that each c_i is a linear combination of the columns of $\mathbf{WK}$, therefore $\mathbf{C} = \mathbf{WKG}$, for some $M \times M$ matrix $\mathbf{G}$. From the orthogonality of c_i it follows

$$\mathbf{I} = \mathbf{C}^T \mathbf{C} = \mathbf{G}^T \mathbf{K}^T \mathbf{WWKG} = \mathbf{G}^T \mathbf{G} \tag{43}$$

that is, the columns of $\mathbf{G}$ form an orthonormal basis of $\mathbb{R}^M$. It is clear that the inverse is also true, more precisely, if the columns of $\mathbf{G}$ are an orthonormal basis of $\mathbb{R}^M$, then the columns of $\mathbf{WKG}$ are an orthonormal basis of $\mathcal{V}_C$.

Moreover, if $B = \cup_i C_i$, a vector f of $\mathcal{V}_B$ can always be obtained by "stacking" vectors of $\mathcal{V}_{C_i}$. The resulting structure of f is, therefore, $f = \mathcal{WK}g$, where $\mathcal{K}$ and $\mathcal{W}$ are defined as

$$\mathcal{K} \triangleq \mathrm{diag}\, \mathbf{K}_1, \ldots, \mathbf{K}_i, \ldots$$

$$\mathcal{W} \triangleq \mathrm{diag}\, \mathbf{W}_1, \ldots, \mathbf{W}_i, \ldots \tag{44}$$

where $\mathbf{K}_i$ and $\mathbf{W}_i$ are, respectively, matrices $\mathbf{K}$ and $\mathbf{W}$ relative to the vector space $\mathcal{V}_{C_i}$. It is also clear that for (44) one can write $(\mathcal{WK})^T \mathcal{WK} = \mathbf{I}$; therefore, all arguments used for the construction of a basis for $\mathcal{V}_C$ still hold.

This way of constructing an orthonormal basis of $\mathcal{V}_B$ can be easily seen as the general extension of the technique to extend a basis by symmetry or antisymmetry, as presented in [12].

The general structure of an orthonormal basis of $\mathcal{V}_B$ is, therefore,

$$\mathcal{W}\mathcal{K}\mathbf{G} \tag{45}$$

where $\mathbf{G}$ is a unitary matrix, "starting basis", and $\mathcal{W}$ is a diagonal matrix having as elements the values that window w assumes on B. Note that $\mathrm{Tr}\mathcal{W}^2 = \mathrm{const}$, that is, the sum of all the elements of $\mathcal{W}$ squared ($\mathcal{W}$ is diagonal) equals a constant.

It is interesting to see how the coefficients of a given function $f \in \mathcal{V}_X$ can be computed. First, observe that $\mathcal{V}_B$ is a subspace of $\chi_B \mathcal{V}_X$, therefore, we can first project f on $\chi_B \mathcal{V}_X$, by obtaining a vector with $|B|$ components; note that such a step does not involve any operations. Next, multiply $\chi_B f$ by $(\mathcal{W}\mathcal{K}\mathbf{G})^T = \mathbf{G}^T \mathcal{K}^T \mathcal{W}$ to obtain the components of f. Since $\mathbf{W}$ is a $|\tilde{B}| \times |\tilde{B}|$ diagonal matrix, the first product costs only $|\tilde{B}|$ products, therefore it has linear complexity. The cost of the product by $\mathcal{K}$ depends on the symmetries.

Suppose, for simplicity's sake, that, for each B_i and each $x \in B_i$ the only $u \in \Gamma$ such that $u(x) = x$ is $u = \mathcal{I}$. Note that in this case $\mathrm{Stab}(x) = \{\mathcal{I}\}$. Such a fact implies that each orbit of B_i has $|\Gamma|$ points and that $|\tilde{B}_i| = |B_i|/|\Gamma|$. Therefore, the block of $\mathcal{K}$ corresponding to B_i will have $|B_i|/|\Gamma|$ rows and $|\Gamma|$ columns with values ± 1 and will cost $(|B_i|/|\Gamma|)(|\Gamma| - 1) \approx |B_i|$ sums. The total cost of $\mathcal{K}$ will be, therefore, less then $|B|$, i.e., linear with the number of point of B.

The product with $\mathbf{G}^T$, instead, will cost in general approximately $2|\tilde{B}|^2$ operations, that is a quadratic cost that can be responsible for the greatest part of computational weight. If we want an orthonormal basis that allows for a fast algorithm we must choose a matrix $\mathbf{G}$ that can be computed with a small number of operations. Some examples of such fast orthonormal bases can be the DCT, the DFT (although it requires complex arithmetic), the discrete Hartley transform (DHT) or the Walsh transform. It is worth noting that a fast algorithm for $\mathbf{G}$ can be immediately transformed into a fast algorithm for the local orthogonal system.

Let us observe that in a filter-bank interpretation of this system, the columns of matrix (45) are the impulse responses of the filters.

3.4. An Orthonormal Basis: General Case

We presented a general technique to find an orthonormal basis for $\mathcal{V}_{B_i}$, when B_i has a finite number of points. We did that by dividing B_i into appropriate $C_{i,j}$ and then found an orthonormal basis for each $\mathcal{V}_C(\Gamma, s, w)$, where $C = C_{i,j}$. For convenience, in what follows, $B = B_i$. In this section we will present how to construct an orthonormal basis in the general case. The scenario will be the following: we have a vector space $\mathcal{V}_B$ that can be written as the direct sum of some subspaces $\mathcal{V}_{C_i}(\Gamma_i, s_i, w_i)$, that is, $\mathcal{V}_B = \oplus_i \mathcal{V}_{C_i}(\Gamma_i, s_i, w_i)$. Functions of the vector space $\mathcal{V}_{C_i}(\Gamma_i, s_i, w_i)$ are "constrained" since they are of the form $f = w_i S_i$, with $S_i(u(x)) = s_i(u)S_i(x)$. Indeed, in the previous section we expressed the vectors of $\mathcal{V}_B$ as linear transformations of vectors of a "free" vectors space for which it is easy to find an orthonormal basis. This section is subdivided in two subsections: in the first we will show a technique to figure out an orthonormal basis of $\mathcal{V}_{C_i}(\Gamma_i, s_i, w_i)$ by using an orthonormal basis of a "free" vector space. This allows us to construct a basis for $\mathcal{V}_B$ by addition. However, in the second subsection, we will show that there exists a more direct method, similar to that explained in the first subsection.

A Basis for $\mathcal{V}_{C_i}(\Gamma_i, s_i, w_i)$

For notational convenience, let us drop the subscript from $\mathcal{V}_{C_i}(\Gamma_i, s_i, w_i)$ and let $\mathcal{V}_C(\Gamma, s, w)$ be the vector space for which we are searching the basis. The main step will be the decomposition of set C into a disjoint union of some opportune subsets.

In the discrete-time case orbits played a central role. Indeed, we could choose one point per orbit to obtain the support of a "free" vector space $\mathbb{R}^N$. Unfortunately, in the continuous-time case we have a nonnumerable number of orbits and the discrete-time approach cannot be used directly. However, we can resort to the orbit approach by considering "orbits of sets". For example, in the one-dimensional continuous-time case, if $C = [0, 2\ell]$ and u is the symmetry around ℓ, it is intuitive that $\tilde{C} \overset{\Delta}{=} [\ell, 2\ell]$ could play a role analogous to the set of representatives in the discrete-time case. More precisely, we will chose a collection $\mathcal{D} \overset{\Delta}{=} \{D_1, \ldots, D_{N_B}\}$ of subsets of C such that C can be reconstructed by applying the symmetries of Γ to the sets of $\mathcal{D}$ or, more formally, such that $C = \cup_i \cup_{u \in \Gamma} u(D_i)$. For theoretical reasons, it will be useful that sets D_i have some properties that can be loosely described as a "good behavior" with respect to the symmetries of Γ. Such a "good behavior" is formalized in the following definition:

DEFINITION 5 *Let* $\mathcal{D} \overset{\Delta}{=} \{D_1, \ldots, D_{N_B}\}$ *be a collection of subsets of* C. *We will say that* $\mathcal{D}$ *is a* valid decomposition *of* C *if*

1. *For each* $D_i \in \mathcal{D}$, *all points of* D_i *share the same stabilizer, that is,* $\forall x, y \in D_i$, $Stab(x) = Stab(y)$. *We will denote the common stabilizer as* $Stab(D_i)$.

2. *For* $u \in \Gamma$ *and* $i \neq j$, $u(D_i) \cap D_j = \emptyset$.

3. *If* $D_i \cap u(D_i) \neq \emptyset$, *then* $u \in Stab(D_i)$.

4. *Set* C *can be written as*

$$C = \cup_{i=1}^{N_B} \cup_{u \in \Gamma} u(D_i). \tag{46}$$

It is intuitive that sets D_i will work like the orbit representatives in the discrete-time case. With this point of view, the counterpart of the set of representatives will be the *reduction of* C, defined as follows:

DEFINITION 6 *Let* $\mathcal{D} \overset{\Delta}{=} \{D_1, \ldots, D_{N_B}\}$ *be a valid decomposition of* C. *The* reduction *of* C *will be defined as*

$$\tilde{C} = \cup_{i=1}^{N_B} D_i. \tag{47}$$

The fact that $\tilde{C}$ is the counterpart of the set of representatives of the discrete-time case is confirmed by Property 6 given below.

It is worth making the following observations about Definition 5:

1. Note that $\text{Stab}(D_i)$ is not the set G_{D_i} of transformations $u \in \Gamma$ that map D_i in itself, but a subset of it. See below for an example where $\text{Stab}(D_i)$ is a proper subset of G_{D_i}.

2. The second condition is equivalent to requiring that $u(D_i) \cap v(D_j) = \emptyset$ when $i \neq j$. This can be readily verified since $u(D_i) \cap v(D_j) = u(D_i \cap u(v(D_j))) = \emptyset$ if and only if $D_i \cap u(v(D_j)) = \emptyset$.

3. The third condition is equivalent to requiring that $u(D_i) \cap v(D_i) \neq \emptyset$ if and only if $uv^{-1} = uv \in \text{Stab}(D_i)$. The proof is similar to that for the second condition.

Note that the second, third, and fourth conditions imply that collection $\{u(D_i) \mid u \in \Gamma, D_i \in \mathcal{D}\}$ is a partition of C. As in the discrete-time case, we will keep only sets D_i such that $s(u) = 1$ for each $u \in \text{Stab}(D_i)$, since, if there exists $u \in \text{Stab}(D_i)$ such that $s(u) = -1$, then, for each $f \in \mathcal{V}_C(\Gamma, s, w)$, $\chi_D, f = 0$. For a valid decomposition we have the following properties:

PROPERTY 6 *If sets $\mathcal{D} = \{D_1, \ldots, D_{N_B}\}$ form a valid decomposition, then $\tilde{C}$ has one and only one representative from each orbit resulting from the action of Γ on set C.*

PROPERTY 7 *If sets $\mathcal{D} = \{D_1, \ldots, D_{N_B}\}$ form a valid decomposition then, for each i, another valid decomposition can be obtained by replacing $x \in D_i$ with $u(x)$, $u \in \Gamma$.*

It is worth looking at some examples. Let us begin with the one-dimensional discrete-time case. The set C is the support of a "tail" of the window and we can suppose that it be the set $0 \leq n \leq N - 1$ and that the symmetry is the function $u[n] = (N - 1) - n$. Note that if N is even we do not have any fixed point (that is, a point such that $u[n] = n$), while if N is odd we have $u[(N - 1)/2] = (N - 1)/2$. By looking at Figure 4(a), where $N = 8$ one can see that we have a valid decomposition with only one set, namely $D_1 = \{0, \ldots, 3\}$, with $\text{Stab}(D_1) = \{\mathcal{I}\}$, because there are no fixed points. In Figure 4(b) we have $N = 9$ and $\mathcal{D}$ must contain at least two subsets, for example $D_1 = \{0, \ldots, 3\}$ and $D_2 = \{4\}$. Note that $\text{Stab}(D_1) = \{\mathcal{I}\}$ and that $\text{Stab}(D_2) = \{\mathcal{I}, u\} = \Gamma$. It is easy to verify all the required properties:

1. We just verified the first property both for D_1 and D_2.

2. It is clear that $D_1 \cap D_2 = \emptyset$ and $u(D_1) \cap D_2 = \emptyset$.

3. The third property is clearly verified.

4. We have $\{0, \ldots, 8\} = D_1 \cup D_2 \cup u(D_1)$.

Note that we could not have chosen $\mathcal{D} = \{D_1\}$ with $D_1 = \{0, 1, 2, 3, 4\}$, because the first property would not have be verified. Set $\tilde{C}$ is $\{0, \ldots, 4\}$ and it contains one and only one representative for each orbit of the action of Γ on C, as already stated by Property 6.

Now let us consider the two-dimensional continuous-time case. In this case B can be the square $[-2\ell \ldots 2\ell] \times [-2\ell \ldots 2\ell]$, with edge length equal to 4ℓ, and C can be one of its four "corners" with $\Gamma = \{\mathcal{I}, u_x, u_y, u_{xy}\}$ being the group of the symmetries around the two axes of C. We can consider C to be $C = C_1 = [-2\ell \ldots 0] \times [-2\ell \ldots 0]$, the lower left corner of B (see Figure 5). Note that in the continuous-time case (both one or two-dimensional) we always have some fixed point; more precisely, the points of x-axis without the origin have stabilizer $\{\mathcal{I}, u_y\}$, the point of y-axis without the origin have stabilizer $\{\mathcal{I}, u_x\}$ and the origin has stabilizer $\{\mathcal{I}, u_x, u_y, u_{xy}\} = \Gamma$. Each other point of C has the trivial stabilizer $\{\mathcal{I}\}$. An opportune choice of sets D_i can be

1. Set D_1 is the semi-open square $(-\ell, 0] \times (-\ell, 0]$. We have $\mathrm{Stab}(D_1) = \{\mathcal{I}\}$.

2. Sets D_2 and D_3 are, respectively, $D_2 = (-\ell, 0] \times [-\ell]$ and $D_3 = [-\ell] \times (-\ell, 0]$. The stabilizers of D_2 and D_3 are $\{\mathcal{I}, u_y\}$ and $\{\mathcal{I}, u_x\}$.

3. Finally, set D_4 contains only the point $(-\ell, -\ell)$ and its stabilizer is the whole group Γ.

It is worth noting again that $\tilde{C} = D_1 \cup \ldots \cup D_4$ has one and only one representative per each orbit. It is clear that a function from $\mathcal{V}_C(\Gamma, s, w)$ is uniquely determined by its value on $\tilde{C}$ and that such values are "free", that is, they are independent of one another. Another useful observation is that we cannot choose, for example, $D_2 = [-2\ell, 0] \times [-\ell]$, because it would not satisfy the first condition of Definition 5, and we cannot choose $D_2 = [-2\ell, -\ell) \times [-\ell] \cup (-\ell, 0] \times [-\ell]$ (the previous D_2, but without the point $(-\ell, -\ell)$) because $u_y(D_2) \cup D_2 = D_2 \neq \emptyset$, but $u_y \notin \mathrm{Stab}(D_2)$ and the third condition is not met. Note that this is a case with $G_{D_2} = \Gamma \neq \mathrm{Stab}(D_2)$. Moreover, with such a definition of D_2, $\tilde{C}$ would have *two* representatives for some orbit and the values assumed by each $f \in c\mathcal{V}_C(\Gamma, s, w)$ on $\tilde{C}$ would not be independent.

The two-dimensional discrete-time case is entirely analogous to the continuous-time one. The only subcases are when the edges have an even or an odd number of points and we will not further elaborate on it.

We saw that each function of $\mathcal{V}_C(\Gamma, s, w)$ is uniquely determined by its values on $\tilde{C}$; it is intuitively appealing to think that a basis of $\mathcal{V}_C(\Gamma, s, w)$ could be obtained from a basis of $\chi_{\tilde{C}} \mathcal{V}_X$.

To this end, we introduce the *redundancy function* on C, defined as

$$R_\Gamma(x) = \sum_{u \in \Gamma} \chi_{u(\tilde{C})}(x). \tag{48}$$

The redundancy function enjoys certain properties given by Lemma 1 in Appendix C. We only need for now that for each $u \in \Gamma$, $R_\Gamma(x) = |\mathrm{Stab}(x)|$, which tell us how many times x is obtained from $u(x)$, when u scans Γ.

Our goal is to construct a basis for $\mathcal{V}_C(\Gamma, s, w)$. Since $\chi_{\tilde{C}} \mathcal{V}_X$ is "free" (that is, it is the vector space of the functions of $\mathcal{V}_X$ having support $\chi_{\tilde{C}}$ without particular constraints) it would be fine if we were able to obtain an orthonormal basis of $\mathcal{V}_C(\Gamma, s, w)$ from an orthonormal basis of $\chi_{\tilde{C}} \mathcal{V}_X$ since it will be simpler to find the latter. In particular, if $|\tilde{C}|$ is finite, then $\chi_{\tilde{C}} \mathcal{V}_X$ is a subspace of $\mathbb{R}^{|\tilde{C}|}$ and it is immediate to find a basis. To achieve such

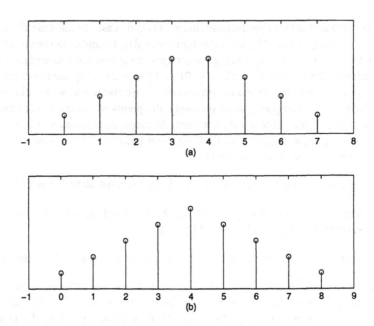

Figure 4. Reduction set (set of representatives) for the one-dimensional discrete-time case. (a) The number of points is even and thus the decomposition contains just one set $D_1 = \{0, 1, 2, 3\}$. (b) The number of points is odd and thus the decomposition contains two sets, for example $D_1 = \{0, 1, 2, 3\}$, $D_2 = \{4\}$.

a goal we need a mapping from $\chi_{\bar{C}}\mathcal{V}_X$ into $\mathcal{V}_C(\Gamma, s, w)$ that preserves the orthonormality. We will use the following linear operator:

$$f = \mathfrak{P}g \stackrel{\Delta}{=} w \sum_{u \in \Gamma} (g/\sqrt{R_\Gamma})_u s(u) = P(\Gamma, s, w)(g/w\sqrt{R_\Gamma}). \tag{49}$$

We claim that each function of $\mathcal{V}_C(\Gamma, s, w)$ can be obtained by an opportune function of $\chi_{\bar{C}}\mathcal{V}_X$ via (49) and that (49) maps an orthonormal basis of $\chi_{\bar{C}}\mathcal{V}_X$ into an orthonormal basis of $\mathcal{V}_C(\Gamma, s, w)$. This is formalized in the following two properties proved in Appendix C.

PROPERTY 8 *Function f in (49) is a vector of $\mathcal{V}_C(\Gamma, s, w)$ and each vector of $\mathcal{V}_C(\Gamma, s, w)$ can be written in form (49), that is, operator $\mathfrak{P}$ is invertible. Moreover, the inverse of $\mathfrak{P}$ has form*

$$\mathfrak{P}^{-1} f = \frac{\chi_{\bar{C}} S}{\sqrt{R_\Gamma}} \tag{50}$$

where S is the function such that $f = wS$ and $S_u = s(u)S$.

PROPERTY 9 *Let $f, g \in \chi_{\bar{C}}\mathcal{V}_X$, then*

$$\langle f, g \rangle = \langle \mathfrak{P}f, \mathfrak{P}g \rangle. \tag{51}$$

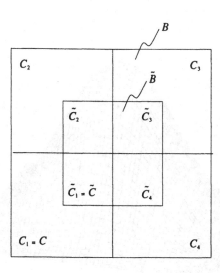

Figure 5. Reduction set for the two-dimensional continuous-time case.

By using the previous two properties it is easy to construct a basis for $\mathcal{V}_C(\Gamma, s, w)$ starting from a basis of $\chi_{\tilde{C}}\mathcal{V}_X$. Indeed, we have the following property:

PROPERTY 10 *Let* g_i, $i = 1, 2, \ldots$, *be an orthonormal basis for* $\chi_{\tilde{C}}\mathcal{V}_X$, *then* $f_i \stackrel{\triangle}{=} \mathfrak{P}_{g_i}$, $i = 1, 2, \ldots$, *is an orthonormal basis for* $\mathcal{V}_C(\Gamma, s, w)$.

Therefore, by using Property 10, one can choose an orthonormal basis of $\chi_{\tilde{C}}\mathcal{V}_X$ and the corresponding orthonormal basis of $\mathcal{V}_C(\Gamma, s, w)$ is obtained by applying $\mathfrak{P}$ to each vector of the basis of $\chi_{\tilde{C}}\mathcal{V}_X$.

As an example, consider $C = [-2\ell \ldots 0] \times [-2\ell \ldots 0]$ and $\tilde{C} = [-\ell \ldots 0] \times [-\ell \ldots 0]$, as in Figure 5. The window relative to B is shown in Figure 6(b) and the "corner" relative to C is shown in Figure 6(b). In Figure 7 one can see some of the functions of an orthonormal basis of $\chi_{\tilde{C}}\mathcal{V}_X$: they are trigonometric functions periodic on the square $\tilde{C}$. In Figure 8 we show the application of operator $\mathfrak{P}$ from (49) to one function from Figure 7. More precisely, in Figure 8(a) one can see the original function; in Figures 8(b) to (e) are shown the results of the application of the symmetries in Γ to the function in (a). In Figures8(f) to (i) one can see the the functions in (b)–(e) multiplied by $s(u)w$, where u is the symmetry applied to (a) to obtain the corresponding figure. Finally, the functions in Figures8(f) to (i) are summed together in order to obtain function in Figure 8(j). If such a procedure is applied to every function in Figure 7 one obtains the functions shown in Figure 9.

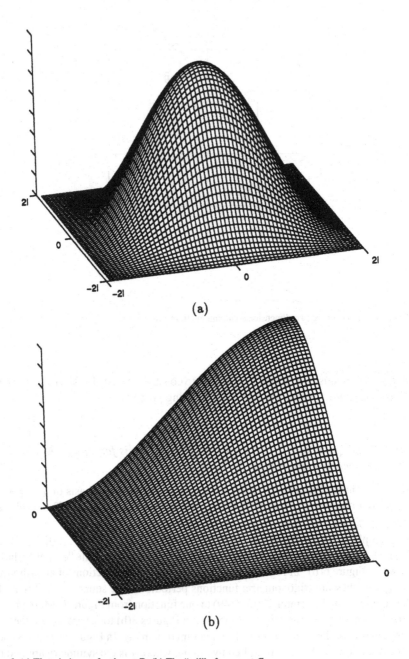

(a)

(b)

Figure 6. (a) The window w for the set B. (b) The "tail" of w on set C.

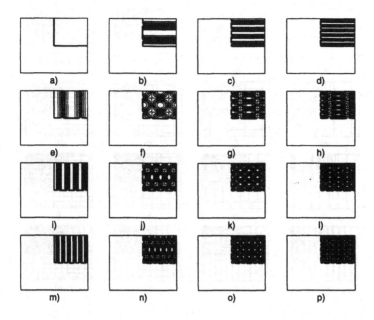

Figure 7. Some of the basis functions from an orthonormal basis of $\chi_{\tilde{C}}\mathcal{V}_X$.

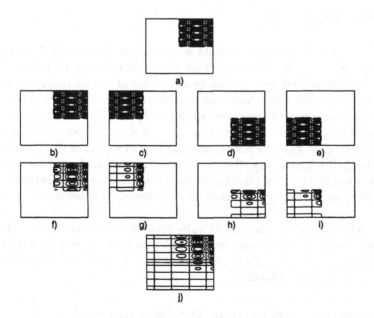

Figure 8. Applying $\mathfrak{P}$ of (49) to one of the basis functions from Figure 7. (a) Original function f. (b) $f_{\mathcal{I}}$. (c) f_{u_y}. (d) f_{u_x}. (e) $f_{u_{xy}}$. (f)–(i) $f_u s(u)w$. (j) Sum of the functions in parts (f)–(i).

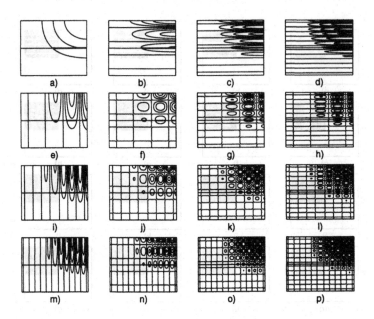

Figure 9. Resulting basis functions obtained from the basis functions for $\chi_{\tilde{C}} \mathcal{V}_X$ in Figure 7 by applying (49).

An Orthonormal Basis for $\mathcal{V}_B$

Now we have a tool that allows us to build an orthonormal basis for $\mathcal{V}_C(\Gamma, s, w)$ from an orthonormal basis of $\chi_{\tilde{C}} \mathcal{V}_X$. However, in the applications $\mathcal{V}_C(\Gamma, s, w)$ is only a building block of a larger vector space $\mathcal{V}_B$ made by direct sum of several vector spaces such as $\mathcal{V}_C(\Gamma, s, w)$. Since the sum is direct an orthonormal basis of $\mathcal{V}_B$ can be obtained by summing up the orthonormal bases of the smaller subspaces. However, there exists a more "direct" method. The idea is that, since $B = \cup_i C_i$ and $\tilde{B} = \cup_i \tilde{C}_i$, we can obtain a basis of $\mathcal{V}_B$ from a basis g_j of $\chi_{\tilde{B}} \mathcal{V}_X$ by first constructing a set of vectors of $\tilde{C}_i$ by multiplying g_j by the characteristic function $\chi_{\tilde{C}_i}$, then these vectors can be transformed via $\mathfrak{P}$ in order to obtain vectors of $\mathcal{V}_{C_i}(\Gamma_i, s_i, w_i)$ that can be recombined in order to obtain (maybe) a basis of $\oplus \mathcal{V}_{C_i}(\Gamma_i, s_i, w_i) = \mathcal{V}_B$. In more detail, the procedure is the following:

1. First "split" a vector of the basis of $\chi_{\tilde{B}} \mathcal{V}_X$ into the sum of vectors of $\chi_{\tilde{C}_i} \mathcal{V}_X$. More precisely, if g_j is an orthonormal basis of $\chi_{\tilde{B}} \mathcal{V}_X$ compute $f_{j,i} = g_j \chi_{\tilde{C}_i} \in \chi_{\tilde{C}_i} \mathcal{V}_X$. Note that system $f_{j,i}$ generates $X_{\tilde{C}_i} \mathcal{V}_X$, although it is not granted to be a basis, since some $f_{j,i}$ could be linearly dependent.

2. To each vector $f_{i,j}$ apply the operator $\mathfrak{P}_i$ to obtain $h_{j,i}$, that is

$$h_{j,i} = \mathfrak{P}_i f_{j,i} = \mathfrak{P}_i(g_j \chi_{\tilde{C}_i}). \tag{52}$$

3. "Mount" vectors $h_{j,i}$ in order to obtain

$$k_j = \sum_i h_{j,i} = \sum_i \mathfrak{P}_i(g_j \chi_{\tilde{C}_i}). \tag{53}$$

We claim that k_j are an orthonormal basis for V_B, that is, they are orthonormal and generate V_B. We will denote the operator defined by (53) as $\mathfrak{Q}$. With such a notation (53) becomes

$$k_j = \mathfrak{Q}g_j. \tag{54}$$

The proof of our claim relies on two properties analogous to Properties 8 and 9 of the preceding subsection. The proofs of these properties are given in Appendix C.

PROPERTY 11 *Let* $g \in \chi_{\tilde{B}} V_X$, *then* $\mathfrak{Q}g \in V_B$. *Moreover, each vector of* V_B *can be written as* $\mathfrak{Q}g$ *with* $g \in V_B$, *that is,* $\mathfrak{Q}$ *is invertible.*

PROPERTY 12 *Let* $g_1, g_2 \in \chi_{\tilde{B}} V_X$, *then*

$$\langle g_1, g_2 \rangle = \langle \mathfrak{Q}g_1, \mathfrak{Q}g_2 \rangle. \tag{55}$$

It is clear that this is the same situation as in the previous subsection, that is

PROPERTY 13 *Let* g_i *an orthonormal basis for* $\chi_{\tilde{B}} V_X$, *then* $f_i \overset{\Delta}{=} \mathfrak{Q}g_i$ *is an orthonormal basis for* V_B.

Consider the following example: Let $B = [-2\ell \ldots 2\ell] \times [-2\ell \ldots 2\ell]$, and, as usual, $B = C_1 \cup C_2 \cup C_3 \cup C_4$, with $C_1 = [-2\ell \ldots 0] \times [-2\ell \ldots 0]$, $C_2 = [-2\ell \ldots 0] \times [0 \ldots 2\ell]$, $C_3 = [0 \ldots 2\ell] \times [-2\ell \ldots 0]$, $C_4 = [0 \ldots 2\ell] \times [0 \ldots 2\ell]$, the four "corners" of B. In this case $\tilde{C}_1 = [-\ell \ldots 0] \times [-\ell \ldots 0]$ and similarly for others $\tilde{C}_i$. The resulting $\tilde{B}$ is $\tilde{B} = [-\ell \ldots \ell] \times [-\ell \ldots \ell]$ (see Figure 5).

Figure 6(a) shows the window relative to B, in Figure 10 one can see a basis for $\chi_{\tilde{B}} V_X$ and in Figure 11(a) we show an element of such a basis. Figures 11(b) to (e) show the result of the multiplication of the function of Figure 11(a) by $\chi_{\tilde{C}_i}$. To the functions of (b) to (e) the corresponding operator $\mathfrak{P}_i$ is applied and the results are shown in Figures 11(f) to (i). Figure 11(j) shows the sum of function of (f) to (i). Finally, in Figure 12 we show the result of applying $\mathfrak{Q}$ to each function of Figure 10.

3.5. Summary

In conclusion, an algorithmic procedure to obtain the decomposition (22) can be the following:

- Choose sets B_i. In the one-dimensional continuous-time case $B_i = [\alpha_i - \epsilon_i, \alpha_{i+1} + \epsilon_{i+1}]$.

- Decompose each B_i as $B_i = \cup_k C_{i,k}$, such that $C_{i_1,k_1} \cap C_{i_2,k_2} = \emptyset$ and that for each $C_{i,k}$ and B_j either $C_{i,k} \subset B_j$ or $C_{i,k} \cap B_j = \emptyset$. In one dimension we have the two transition zones $C_{i,-1} = [\alpha_i - \epsilon_i, \alpha_i + \epsilon_i]$, $C_{i,1} = [\alpha_i + \epsilon_i, \alpha_{i+1} + \epsilon_{i+1}]$ and the central zone $C_{i,0} = [\alpha_i + \epsilon_i, \alpha_{i+1} - \epsilon_{i+1}]$.

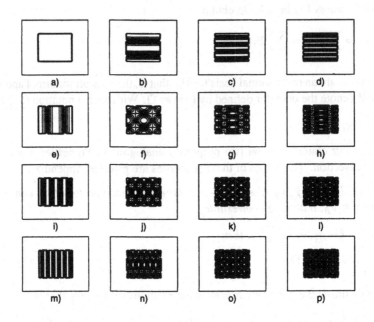

Figure 10. Some of the basis functions from an orthonormal basis of $\chi_{\tilde{B}} \mathcal{V}_X$.

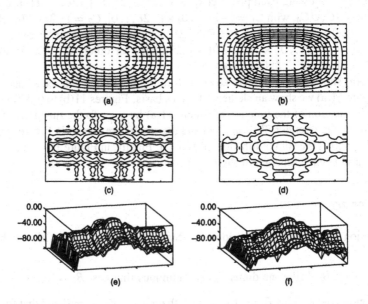

Figure 11. Applying Ω to one of the basis functions from Figure 10. (a) Original function f. (b) $\chi_{\tilde{C}_2} f$, (c) $\chi_{\tilde{C}_3} f$, (d) $\chi_{\tilde{C}_1} f$, (e) $\chi_{\tilde{C}_4} f$. (f)–(i) $\mathfrak{P}_i f$. (j) Sum of the functions in parts (f)–(i).

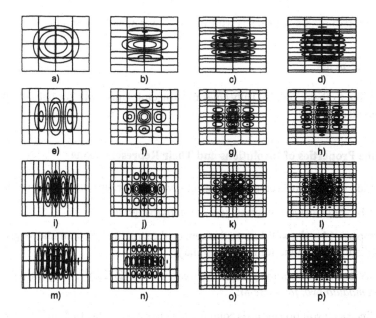

Figure 12. Resulting basis functions obtained from the basis functions for $\chi_{\tilde{B}}\mathcal{V}_X$ in Figure 10 by applying (53).

- For each $C_{i,k}$ find all B_j's such that $C_{i,k} \subset B_j$. For each B_j there will be a vector space $\mathcal{V}_j$. Choose symmetries, window and weights such that the sum of $\sum_j \mathcal{V}_j$ is direct and it is equal to $\chi_{C_{i,k}}\mathcal{V}_X$. To this end Properties 1–4 can be useful.

  It is worth noting that if some $C_{i,k}$ is contained only in one B_j (for example $[\alpha_i + \epsilon_i, \alpha_{i+1} - \epsilon_{i+1}]$ is contained only in $[\alpha_i - \epsilon_i, \alpha_{i+1} + \epsilon_{i+1}]$) it must be $\mathcal{V}_j = \chi_{C_{i,k}}\mathcal{V}_X$. This can be seen as a particular case of spaces described in Definition 1 with $\Gamma = \{\mathcal{I}\}$ and $w = \chi_{C_{i,k}}$, as required by condition of normalization of powers in Definition 1.

- Define $\mathcal{V}_{B_i} \overset{\Delta}{=} \oplus_j \mathcal{V}_{c_{ij}}$, where $\mathcal{V}_{C_{i,j}}(\Gamma, s, w)$ is the vector space with support $C_{i,j}$. These spaces, by construction, satisfy (22).

- Choose a basis for $\mathcal{V}_{B_i}$. For example, in discrete time, choose the starting basis **G** (for example, it could be a DCT) and form the final basis $\mathcal{WKG}$, where $\mathcal{W}$ and $\mathcal{K}$ are obtained from the window and weights, respectively.

4. Conclusions

This work presented a general extension of the one-dimensional local cosine bases using the subdivision of a vector space into a direct sum of some "local" spaces. The theory presented is very general and can handle a large variety of cases interesting for applications, ranging from the continuous to the discrete time, from the one-dimensional case to the multidimensional one.

An integral part of this work deals with window design, and is given in [13].

Acknowledgements

We would like to thank Ingrid Daubechies of Princeton University for helpful comments and fruitful discussions.

A   Some Properties of Involutions and Their Representations

In this section we will give some properties of involution groups used in the work.

PROPERTY 14 *Each involution group is a commutative group.*

From Property 14 and a well-known theorem on the structure of commutative groups [11] it is immediate to obtain the following corollary:

COROLLARY 1 *Each involution group Γ is isomorphic to $(\mathbb{Z}/2\mathbb{Z})^M$, the group of M-tuples of integers modulo two, for some M.*

Proof: Since an involution group is commutative, it is isomorphic to $\mathbb{Z}/a_1\mathbb{Z} \times \ldots \times \mathbb{Z}/a_M\mathbb{Z}$ [11]. Since each element of Γ has period two it follows $a_1 = \ldots a_M = 2$. ∎

In group theory the concept of representations is often used.

DEFINITION 7 *Let Γ be a group, V a vector space and $GL(V)$ the group of nonsingular linear transformation on V. A representation s of Γ is a homomorphism mapping Γ to $GL(V)$.*

In particular, a homomorphism between Γ and $\mathbb{C}$ is called a *representation of degree 1*.

If Γ is an involution group (enjoying, consequently, the structural property of Corollary 1), then each representation of Γ can be constructed by assigning a "weight" $s(u_i) = \pm 1$ to each element of the basis (or, equivalently, to each bit b_i) and by defining s on the other elements by "linearity", that is, if $u = u_1 \ldots u_M$, where u_i are involutions of the basis, define $s(u) \overset{\Delta}{=} s(u_1) \ldots s(u_M)$. Note that for each representation $s(\mathcal{I}) = 1$.

If Γ_1 and Γ_2 are two groups of involutions, we will denote with $< \Gamma_1, \Gamma_2 >$ the group generated by their combination, that is

$$< \Gamma_1, \Gamma_2 > \overset{\Delta}{=} \{\gamma : \gamma = u_1 u_2; \quad u_i \in \Gamma_1\}. \tag{56}$$

Note that each $u_1 \in \Gamma_1$ belongs also to $< \Gamma_1, \Gamma_2 >$, since $u_1 = u_1\mathcal{I}$. This implies that Γ_1 is a subgroup of $< \Gamma_1, \Gamma_2 >$. A similar property holds for $\Gamma[2]$.

If Γ_1 and Γ_2 are two involution groups and t, r are their representations, it is easy to show that a representation s_3 for $< \Gamma_1, \Gamma_2 >$ can be constructed as

$$s_3(u_1 u_2) = t(u_1) r(u_2), \quad u_i \in \Gamma_i \tag{57}$$

Construction (57) of s_3 from r and t will be denoted as $s_3 = t \otimes r$.

PROPERTY 15 *If* $\Gamma = < \Gamma_1, \Gamma_2 >$ *and if s is a representation of* Γ, *then it is possible to find representations t and r of* Γ_1 *and* Γ_2, *such that*

$$s = t \otimes r. \tag{58}$$

Property 15 allows us to split each representation of Γ, if we can split Γ. Note that by thinking of Γ as described in Corollary 1, to split an involution group corresponds to splitting the M-bit number into two smaller fields, if the basis is adequately chosen. The proof corresponds to "giving" the weights of the elements of the basis of Γ_1 to t and similarly for Γ_2 and r.

B   Projection Operators and their Properties

In this section we will give a brief overview of projection operators and properties that we use in the derivations. For proofs and more details, see, for example [15].

DEFINITION 8 *Let S be a closed subspace of a Hilbert space H. The operator P on H defined by*

$$P(x) = y,$$

for $x = y + z$, *where* $y \in S$ *and* $z \in S^{\perp}$, *is called a projection operator onto S. The vector y is called the projection of x onto S.*

DEFINITION 9 *An operator P is called idempotent if* $P^2 = P$.

DEFINITION 10 *An operator P is called self-adjoint if* $< Px, y > = < x, Py >$ *for all x, y in the Hilbert space H.*

PROPERTY 16 *A bounded operator is a projection if and only if it is idempotent and self-adjoint.*

DEFINITION 11 *Two projection operators* P_1 *and* P_2 *are orthogonal if* $P_1 P_2 = 0$.

PROPERTY 17 *Two projection operators* $P_1 : H \to S_1$ *and* $P_2 : H \to S_2$ *are orthogonal if and only if* $S_1 \perp S_2$.

PROPERTY 18 *The sum of two projection operators* $P_1 : H \to S_1$ *and* $P_2 : H \to S_2$ *is a projection operator if and only if* $P_1 P_2 = 0$. *In this case* $P_1 + P_2 = P : H \to S_1 \oplus S_2$.

PROPERTY 19 *The product of two projection operators* $P_1 : H \to S_1$ *and* $P_2 : H \to S_2$ *is a projection operator if and only if* $P_1 P_2 = P_2 P_1$. *In this case* $P_1 P_2 = P : H \to S_1 \cap S_2$.

C   Proofs of Properties

In this section we present proofs of some of the properties reported in this work. First, we will prove that (26) is an orthogonal projection. According to Property 16 this is equivalent to showing that (26) is self-adjoint (that is, $\langle P(\Gamma, s, w)f, g \rangle = \langle f, P(\Gamma, s, w)g \rangle$)) and idempotent (that is, $P(\Gamma, s, w)P(\Gamma, s, w) = P(\Gamma, s, w)$).

Proof:

Step 1: $P(\Gamma, s, w)$ **is self-adjoint:**

$$
\begin{aligned}
\langle P(\Gamma, s, w)f, g \rangle &= \langle \sum_u s(u)ww_u f_u, g \rangle, \\
&= \sum_u s(u)\langle ww_u f_u, g \rangle, \\
&= \sum_u s(u)\langle f, ww_u g_u \rangle, \\
&= \langle f, \sum_u s(u)ww_u g_u \rangle, \\
&= \langle f, P(\gamma, s, w)g \rangle.
\end{aligned}
\tag{59}
$$

It is worth noting that going from the second to the third equation is granted by the two hypotheses that each $u \in \Gamma$ is unitary and that $\langle hf, g \rangle = \langle f, hg \rangle$, when h is real (as w is).

Step 2: $P(\Gamma, s, w)$ **is idempotent:**

To prove that $P(\Gamma, s, w)$ is idempotent it is sufficient to apply it twice to a function $f \in V_X$.

$$
\begin{aligned}
P(\Gamma, s, w)P(\Gamma, s, w)f &= \sum_{u \in \Gamma} s(u)ww_u \left(\sum_{v \in \Gamma} s(v)ww_v f_v \right)_u \\
&= \sum_{u,v \in \Gamma} s(u)s(v)ww_u w_u w_{uv} f_{uv}.
\end{aligned}
\tag{60}
$$

By calling $\gamma = uv$ in (60), one can rewrite

$$
P(\Gamma, s, w)P(\Gamma, s, w)f = \sum_{u \in \Gamma} (w_u)^2 \sum_{\gamma \in \Gamma} s(\gamma)ww_\gamma f_\gamma,
\tag{61}
$$

and, because of power normalization (24), (61) becomes

$$
P(\Gamma, s, w)P(\Gamma, s, w)f = \sum_{\gamma \in \Gamma} s(\gamma)ww_\gamma f_\gamma = P(\gamma, s, w)f.
\tag{62}
$$

∎

Next we prove that $P(\Gamma, s, w)$ is a projection on space $\mathcal{V}_C(\Gamma, s, w)$. We will do that in two steps. First, we will prove that for each $f \in \mathcal{V}_X$, $P(\Gamma, s, w)f \in \mathcal{V}_C(\Gamma, s, w)$. If $P(\Gamma, s, w)f$ is in $\mathcal{V}_C(\Gamma, s, w)$, then it has to be written as wS. We will do that by showing that $P(\Gamma, s, w)f/w$ enjoys symmetries (25). This will grant that $P(\Gamma, s, w)$ is, at least, a projection on some subspace of $\mathcal{V}_C(\Gamma, s, w)$; next, we will verify that $P(\Gamma, s, w)$ generates all $P(\Gamma, s, w)$ by showing that if $f \in \mathcal{V}_C(\Gamma, s, w)$, that is equivalent to $f = wS$, with S satisfying (25), then $P(\Gamma, s, w)f = f$.

Proof:

Step 1: $P(\Gamma, s, w)f/w$ satisfies (25)

First, compute $P(\Gamma, s, w)f/w$.

$$S \stackrel{\Delta}{=} P(\Gamma, s, w)f/w = \sum_{u \in \Gamma} s(u)w_u f_u. \tag{63}$$

Next, compute s_v, $v \in \Gamma$

$$S_v = \sum_{u \in \Gamma} s(u)w_{uv} f_{uv}. \tag{64}$$

In (64), the expression uv spans all group Γ when u varies; therefore one can substitute uv with $t \in \Gamma$ and, consequently, u with tv, obtaining

$$S_v = \sum_{t \in \Gamma} s(tv)w_t f_t. \tag{65}$$

By using the fact that $s(tv) = s(t)s(v)$, (65) can be rewritten as

$$S_v = s(v) \sum_{t \in \Gamma} s(t)w_t f_t = s(v)S. \tag{66}$$

Step 2: $P(\Gamma, s, w)f = f$

Let $f = wS$, with S satisfying (25). By applying $P(\Gamma, s, w)$ to wS one obtains

$$\sum_{u \in \Gamma} s(u)ww_u(wS)_u = \sum_{u \in \Gamma} s(u)ww_u w_u S_u$$
$$= \sum_{u \in \Gamma} s(u)ww_u^2 s(u)S$$
$$= wS \sum_{u \in \Gamma} w_u^2 = wS = f. \tag{67}$$

It is worth noting that in the last step of (67) the condition of power normalization has been used. ∎

Now we can prove Property 1.

Proof of Property 1: Recall from Definition 11 that two projection operators are orthogonal if $P_1 P_2 = 0$. It is sufficient to show that, if $f \in V_C(\Gamma, s, w_v)$, then $P(\Gamma, s, w)f = 0$. Remember that if $f \in V_C(\Gamma, t, w_v)$, then $f = w_v S$, where $S_u = t(u)S$. Because of Corollary 1 it is possible to consider involution group Γ as generated by v and a suitable subgroup Γ_1 of Γ, more precisely, $\Gamma = \{\Gamma_1, v\Gamma_1\}$. Using such a fact one can write

$$
\begin{aligned}
P(\Gamma, s, w)f &= \sum_{u \in \Gamma} s(u)ww_u f_u \\
&= \sum_{u \in \Gamma_1} s(u)ww_u f_u + \sum_{u \in v\Gamma_1} s(u)ww_u f_u \\
&= \sum_{u \in \Gamma_1} s(u)ww_u f_u + \sum_{u \in \Gamma_1} s(v)s(u)ww_{uv} f_{uv} \\
&= \sum_{u \in \Gamma_1} s(u)ww_u w_{uv} t(u)S + \sum_{u \in \Gamma_1} s(v)s(u)ww_{uv} w_u t(u)t(v)S \\
&= \sum_{u \in \Gamma_1} s(u)t(u)ww_u w_{uv} S - \sum_{u \in \Gamma_1} s(u)t(u)ww_{uv} w_u S = 0.
\end{aligned}
\tag{68}
$$

The last step is granted by the hypothesis $s(v) = -t(v)$.   ∎

Proof of Property 3: First, let us prove that if $P_1 \overset{\Delta}{=} P(\Gamma_1, s, w)$ and $P_2 \overset{\Delta}{=} P(\Gamma_2, t, y)$ verify the hypothesis of Property 3, then $P_1 P_2$ is an orthogonal projection. According to Property 19 it is sufficient to show that $P_1 P_2 = P_2 P_1$ [16]. Indeed,

$$
\begin{aligned}
P_1 P_2 f &= \sum_{u \in \Gamma_1} s(u)ww_u \left(\sum_{v \in \Gamma_2} t(v)yy_v f_v \right)_u \\
&= \sum_{u \in \Gamma_1} \sum_{v \in \Gamma_2} s(u)t(v)ww_u yy_u y_{uv} f_{uv} \\
&= \sum_{v \in \Gamma_2} t(v)yy_v \left(\sum_{u \in \Gamma_1} s(u)ww_u f_u \right)_v = P_2 P_1 f.
\end{aligned}
\tag{69}
$$

In the last passage of (69) we used the hypothesis that $w_v = w$ and $y_u = y$, for each $u \in \Gamma_1$ and $v \in \Gamma_2$.

To complete the proof we have to show that $P_1 P_2$ has the form (29). By restarting from the second equation of (69) one obtains

$$
\begin{aligned}
P_1 P_2 f &= \sum_{u \in \Gamma_1} s(u) w w_u \left(\sum_{v \in \Gamma_2} t(v) y y_v f_v \right)_u \\
&= \sum_{u \in \Gamma_1} \sum_{v \in \Gamma_2} s(u) t(v) (wy)(wy)_{uv} f_{uv} \\
&= \sum_{\gamma \in <\Gamma_1, \Gamma_2>} s \otimes t(\gamma)(wy)(wy)_\gamma f_\gamma = P(<\Gamma_1, \Gamma_2>, s \otimes t, wy). \quad (70)
\end{aligned}
$$

∎

Proof of Property 4: We will proceed in three steps: first, it will be proved that θ and τ are two windows satisfying the power-complementarity condition (24) with respect to groups Γ_1 and Γ_2, respectively. In the second step the symmetry conditions (28) required by Property 3 will be verified and finally, in the third step, it will be shown that $w = \theta \tau$, from which, by applying Property 3, Property 4 will follow.

Step 1: Windows θ and τ verify the power-complementarity condition

It is sufficient to verify the condition of power-complementarity for θ, since a similar reasoning can be made for τ. The condition of power-complementarity for θ with respect to Γ_1 is

$$
\sum_{v \in \Gamma_1} \theta_v^2 = \sum_{v \in \Gamma_1} \frac{w_{v^2}}{\sum_{u \in \Gamma_1} w_{uv}^2}. \quad (71)
$$

Since in (71) $v \in \Gamma[1]$, in the sum of the denominator uv spans all Γ_1, that is, $v\Gamma_1 = \Gamma_1$, and $\sum_{u \in \Gamma_1} w_{uv}^2 = \sum_{u \in \Gamma_1} w_u^2$, for each v. Therefore (71) can be rewritten as

$$
\sum_{v \in \Gamma_1} \frac{w_v^2}{\sum_{u \in \Gamma_1} w_u^2} = \frac{\sum_{v \in \Gamma_1} w_v^2}{\sum_{u \in \Gamma_1} w_u^2} = 1. \quad (72)
$$

Step 2: Windows θ and τ verify the symmetry conditions (28)

Let $v \in \Gamma_2$, then

$$
\theta_v = \frac{w_v}{\sqrt{\sum_{u \in \Gamma_1} w_{uv}^2}}. \quad (73)
$$

By multiplying numerator and denominator of (73) by w and using the fact that $w w_{uv} = w_u w_v$ one obtains

$$
\frac{w w_v}{\sqrt{\sum_{u \in \Gamma_1} w^2 w_{uv}^2}} = \frac{w w_v}{\sqrt{\sum_{u \in \Gamma_1} w_u^2 w_v^2}},
$$

$$= \frac{w w_v}{w_v \sqrt{\sum_{u \in \Gamma_1} w_u^2}},$$

$$= \frac{w}{\sqrt{\sum_{u \in \Gamma_1} w_u^2}} = \theta. \tag{74}$$

A similar reasoning can be made for τ.

Step 3: We want to show that $\theta \tau = w$

$$\theta \tau = \frac{w^2}{\sqrt{\sum_{u \in \Gamma_1} \sum_{v \in \Gamma_2} w_u^2 w_v^2}}$$

$$= \frac{w^2}{\sqrt{\sum_{u \in \Gamma_1} \sum_{v \in \Gamma_2} w^2 w_{uv}^2}}$$

$$= \frac{2}{\sqrt{\sum_{\gamma in \Gamma} w_\gamma^2}} = w. \tag{75}$$

To summarize, with θ and τ defined as in (31), satisfying (24), we have all assumptions of Property 3. Thus the conclusion of Property 3 holds, that is, P can be written as (29). ∎

Proof of Property 5: First, observe that if b_i is a basis of $\mathcal{V}_C(\Gamma, s, w_C)$, then $w b_i$ generate all of $\mathcal{V}_C(\Gamma, s, w)$, since each function f of $\mathcal{V}_C(\Gamma, s, w)$ can be written as $f = wS$, where S satisfies symmetries (25), that is, S belongs to $\mathcal{V}_C(\Gamma, s, w_C)$.

Now the orthonormality of functions $\sqrt{|\Gamma|} w b_i$ must be proved. First, we will show the more general fact that $\langle wa, wb \rangle = \langle a, b \rangle / |\Gamma|$, for any $a, b \in \mathcal{V}_C(\Gamma, s, w_C)$.

Indeed, for each $u \in \Gamma$

$$\langle wa, wb \rangle = \langle w^2 a, b \rangle$$

$$= \langle w_u^2 a_u, b_u \rangle$$

$$= \langle w_u^2 s(u)a, s(u)b \rangle$$

$$= \langle w_u^2 a, b \rangle \tag{76}$$

In (76) the second equation is granted by the fact that each $u \in \Gamma$ is supposed unitary by assumption and the third and the fourth follows from $a_u = s(u)a$, $b_u = s(u)b$, since $a, b \in \mathcal{V}_C(\Gamma, s, w_C)$, and $s(u)s(u) = 1$.

By summing with respect to $u \in \Gamma$, the first and the last part of equation (76) one obtains

$$\sum_{u \in \Gamma} \langle wa, wb \rangle = \sum_{u \in \Gamma} \langle w_u^2 a, b \rangle,$$

$$|\Gamma| \langle wa, wb \rangle = \langle \sum_{u \in \Gamma} w_u^2 a, b \rangle,$$

$$\langle wa, wb \rangle = \frac{1}{|\Gamma|} \langle a, b \rangle. \tag{77}$$

Now it is easily seen that

$$\langle w\sqrt{|\Gamma|}b_i, w\sqrt{|\Gamma|}b_j\rangle = \langle\sqrt{|\Gamma|}b_i, \sqrt{|\Gamma|}b_j\rangle/|\Gamma| = 0, \tag{78}$$

for $i \neq j$, and that

$$\langle\sqrt{|\Gamma|}b_i, w\sqrt{|\Gamma|}b_i\rangle = \langle\sqrt{|\Gamma|}b_i, \sqrt{|\Gamma|}b_i\rangle/|\Gamma| = 1, \tag{79}$$

that is, vectors $w\sqrt{|\Gamma|}b_i$ are orthonormal. ∎

Proof of Property 8: The fact that (49) is a function of $\mathcal{V}_C(\Gamma, s, w)$ is obvious, since (49) is the projection on $\mathcal{V}_C(\Gamma, s, w)$ of $g/R_\Gamma w$. In order to complete the proof we have to prove that by applying $\mathfrak{P}$ to (50) one obtains $f = wS$. To verify this, substitute (50) in (49).

$$\sum_{u\in\Gamma} w\left(\frac{\chi_{\tilde{C}}S}{R_\Gamma}\right)_u s(u) = \sum_{u\in\Gamma} w\frac{\chi_{u(\tilde{C})}S_u}{R_\Gamma}s(u)$$

$$= \sum_{u\in\Gamma} wS_u s(u)\frac{\chi_{u(\tilde{C})}}{R_\Gamma} \tag{80}$$

Remember that, since $f \in \mathcal{V}_C(\Gamma, s, w)$, $S_u = s(u)S$. By substituting in (80),

$$\sum_{u\in\Gamma} wS_u s(u)\frac{\chi_{u(\tilde{C})}}{R_\Gamma} = \sum_{u\in\Gamma}\frac{\chi_{u(\tilde{C})}S_u}{R_\Gamma}s(u)$$

$$= \sum_{u\in\Gamma} wSs(u)s(u)\frac{\chi_{u(\tilde{C})}}{R_\Gamma}$$

$$= wS\frac{\sum_{u\in\Gamma}\chi_{u(\tilde{C})}}{R_\Gamma} = wS = f \tag{81}$$

By the definition of R_Γ. ∎

In order to prove Property 9, the following lemma about the properties of the redundancy function $R_\Gamma(x)$, defined in (48), is instrumental.

LEMMA 1 *Let R_Γ be defined as in (48), then*

1. *For each $u \in \Gamma$, $R_\Gamma(u(x)) = R_\Gamma(x)$.*

2. *$R_\Gamma(x) = |Stab(x)|$. Note that this implies that R_Γ is independent on the choice of $\tilde{C}$ and that $R_\Gamma(x) > 0$ since $\mathcal{I} \in Stab(x)$.*

3. *Let p be a function from $\mathbb{R}$ onto $\mathbb{R}$, then*

$$\chi_{D_i}(R_\Gamma) = \chi_{D_i}p(|Stab(D_i)|) \tag{82}$$

Proof of Lemma 1:

Step 1: $R_\Gamma(x) = R_\Gamma(u(x))$
From the definition

$$R_\Gamma(u(x)) = \sum_{v \in \Gamma} \chi_{v(\tilde{C})}(u(x))$$

$$= \sum_{v \in \Gamma} \chi_{\tilde{C}}(v(u(x))) \tag{83}$$

As long as v in (83) scans Γ, uv scans Γ too, but with a different order. Therefore, (83) can be rewritten as

$$R_\Gamma(u(x)) = \sum_{v \in \Gamma} \chi_{\tilde{C}}(v(x)) = R_\Gamma(x) \tag{84}$$

Step 2: $R_\Gamma(x) = |\text{Stab}(x)|$
Since the value of $R_\Gamma(x)$ does not change on the orbit of x, we can choose, without loss of generality, $x \in \tilde{C}$. Let D_i the unique set whom x belongs. Therefore,

$$R_\Gamma(x) = \sum_{u \in \Gamma} \chi_{u(\tilde{C})}(x) = \sum_{u \in \Gamma} \chi_{D_i}(u(x)) \tag{85}$$

Therefore, $R_\Gamma(x)$ is the number of symmetries $u \in \Gamma$ such that $u(x) \in D_i$. From the third condition it follows that $u(x) \in D_i$ if and only if $u \in \text{Stab}(x)$, therefore $R_\Gamma(x) = |\text{Stab}(x)|$.

Step 3: $\chi_{D_i} p(R_\Gamma) = p(|\text{Stab}(D_i)|)\chi_{B_i}$
If $x \notin D_i$, then $\chi_{D_i}(x)p(R_\Gamma(x)) = 0 = p(|\text{Stab}(D_i)|)\chi_{D_i}(x)$ and for each $x \in D_i$ $R_\Gamma(x) = |\text{Stab}(x)| = |\text{Stab}(D_i)|$ and $\chi_{D_i}(x)p(R_\Gamma(x)) = p(|\text{Stab}(D_i)|) = p(|\text{Stab}(D_i)|)\chi_{D_i}(x)$. ∎

Proof of property 9: By substituting the expression of $\mathfrak{P}$ in (51) and by remembering that if $f \in \chi_{\tilde{C}} V_X$, then $f = \chi_{\tilde{C}} f = \sum_i \chi_{D_i} f$, one obtains

$$\left\langle \sum_{u \in \Gamma} w R_\Gamma \left(\sum_i \chi_{D_i} f / \sqrt{R_\Gamma} \right) s(u), \sum_{v \in \Gamma} w R_\Gamma \left(\sum_j \chi_{D_j} g / \sqrt{R_\Gamma} \right) s(v) \right\rangle \tag{86}$$

that, by using the second property of Lemma 1 and bringing out from the inner product sums and multiplications by constants, can be rewritten as

$$\sum_{i,j} \sum_{u,v \in \Gamma} \frac{1}{\sqrt{|\text{Stab}(D_i)||\text{Stab}(D_j)|}} \langle w(\chi_{D_i} f)_u s(u), w(\chi_{D_j} f)_v s(v) \rangle \tag{87}$$

In (87) we have a sum of inner products of functions defined on $u(D_i)$ and $v(D_j)$. The second condition of Definition 5 on sets D_i implies that in (87) remain only the terms for $i = j$ and (87) becomes

$$\sum_i \sum_{u \in \Gamma} \sum_{v \in \Gamma} \frac{1}{|\text{Stab}(D_i)|} \langle w(\chi_{D_i} f)_u s(u), w(\chi_{D_i} g)_v s(v) \rangle \tag{88}$$

In (88) only the terms with $u(D_i) = v(D_i)$, that is, $uv \in \text{Stab}(D_i)$ (see third condition), remain and one can rewrite (88) as

$$\sum_i \sum_{u \in \Gamma} \sum_{v \in \text{Stab}(D_i)} \frac{1}{|\text{Stab}(D_i)|} \langle w(\chi_{D_i} f)_u s(u), w(\chi_{D_i} g)_{uv} s(uv) \rangle. \tag{89}$$

Since $v \in \text{Stab}(D_i)$ it is immediate that $(\chi_{D_i} g)_v = \chi_{D_i} g$. By remembering that $s(uv) = s(u)s(v)$ (89) becomes

$$\sum_i \sum_{u \in \Gamma} \sum_{v \in \text{Stab}(D_i)} \frac{1}{|\text{Stab}(D_i)|} \langle w(\chi_{D_i} f)_u, w(\chi_{D_i} g)_u \rangle s(v)$$

$$= \sum_i \sum_{u \in \Gamma} \langle w^2(\chi_{D_i} f)_u, g_u s(v) \rangle \tag{90}$$

Now, since we decided to keep only sets D_i such that $s(v) = 1$, for each $v \in \text{Stab}(D_i)$ we can drop the term $s(v)$ from the left hand member of (90). Finally, by applying symmetry u to both member of the inner product and by bringing both sums inside the inner product one obtains

$$\langle \sum_i \chi_{D_i} \sum_{u \in \Gamma} w_u^2 f, g \rangle = \langle f, g \rangle \tag{91}$$

because $\sum_{u \in \Gamma} w_u^2 = 1$, for the power-complementarity condition and $\sum_i \chi_{D_i} f = \chi_{\tilde{C}} f = f$, because f has support $\tilde{C}$ and $\tilde{C} = \cup_i D_i$, with $D_i \cap D_j = \emptyset$, for $i \neq j$. ∎

Proof of Property 11: That $\Omega g \in V_B$ is obvious, since it is the sum of projections on $V_{C_i}(\Gamma_i, s_i, w_i)$. We will prove the second part by showing that the inverse of $\Omega = \sum_i \mathfrak{P}_i \chi_{\tilde{C}_i}$ is

$$\Omega^{-1} = \sum_i \mathfrak{P}_i^{-1} \chi_{C_i}, \tag{92}$$

that is, that

$$\Omega \Omega^{-1} h = \sum_i \sum_j \mathfrak{P}_i \chi_{\tilde{C}_i} \mathfrak{P}_j^{-1} \chi_{C_j} h = h, \tag{93}$$

for each $h \in V_B$. Note that in (92) we have the characteristic functions of sets $\mathbf{C}_i$, while in the definition of operator $\mathbf{Q}$ we have the characteristic functions of the reductions $\tilde{C}_i$. Let us observe that applying $\mathbf{P}_j^{-1}$ to $\chi_{C_j} h$ in (93) makes sense, since $h \in \mathcal{V}_B = \oplus_j \mathcal{V}_{C_j}(\Gamma_j, s_j, w_j)$ and $\chi_{C_j} h \in \mathcal{V}_C(\Gamma_j, s_j, w_j)$ that is the domain of $\mathbf{P}_j^{-1}$. In (93) $\mathbf{P}_j^{-1} \chi_{C_j} h$ has support $\tilde{C}_j$ and if $i \neq j$ the corresponding term in (93) is null because $\tilde{C}_j \cap \tilde{C}_i = \emptyset$; therefore (93) simplifies in

$$\mathbf{Q}\mathbf{Q}^{-1} h = \sum_i \mathbf{P}_i \chi_{\tilde{C}_i} \mathbf{P}_i^{-1} \chi_{C_i} h. \tag{94}$$

As already mentioned, $\mathbf{P}_i^{-1} \chi_{C_i} h$ has support $\tilde{C}_i$, therefore we can drop the multiplication by $\chi_{\tilde{C}_i}$ in (94) to obtain

$$
\begin{aligned}
\mathbf{Q}\mathbf{Q}^{-1} h &= \sum_i \mathbf{P}_i \mathbf{P}_i^{-1} \chi_{C_i} h \\
&= \sum_i \chi_{C_i} h = h,
\end{aligned}
\tag{95}
$$

because h belongs to $\mathcal{V}_B$ and has support $B = \cup_i C_i$.   ∎

Proof of Property 12: Let us write $\langle \mathbf{Q} g_1, \mathbf{Q} g_2 \rangle$ by using the definition of $\mathbf{Q}$:

$$\langle \mathbf{Q} g_1, \mathbf{Q} g_2 \rangle = \langle \sum_i \mathbf{P}_i(g_1 \chi_{\tilde{C}_i}), \sum_j \mathbf{P}_j(g_2 \chi_{\tilde{C}_j}) \rangle. \tag{96}$$

In (96) we have sums of inner products between function having supports $\tilde{C}_i$ and $\tilde{C}_j$, respectively. Since $\tilde{C}_i \cap \tilde{C}_j = \emptyset$ if $i \neq j$, in (96) only the terms with $i = j$ remain and (96) can be rewritten as

$$\langle \mathbf{Q} g_1, \mathbf{Q} g_2 \rangle = \sum_i \langle \mathbf{P}_i(g_1 \chi_{\tilde{C}_i}), \mathbf{P}_i(g_2 \chi_{\tilde{C}_i}) \rangle. \tag{97}$$

By Property 9 (97) becomes

$$\langle \mathbf{Q} g_1, \mathbf{Q} g_2 \rangle = \sum_i \langle g_1 \chi_{\tilde{C}_i}, g_2 \chi_{\tilde{C}_i} \rangle = \sum_i \langle g_1 \chi_{\tilde{C}_i}, g_2 \rangle. \tag{98}$$

By bringing the sum inside the inner product one obtains

$$\langle \mathbf{Q} g_1, \mathbf{Q} g_2 \rangle = \langle \sum_i \chi_{\tilde{C}_i} g_1, g_2 \rangle = \langle g_1, g_2 \rangle. \tag{99}$$

∎

Notes

1. These constructions are known under various names: MDCT, LOT, modulated lapped transforms, time-domain aliasing cancelation filter banks, Princen and Bradley filter banks, Malvar wavelets.

References

1. H. Malvar, *Signal Processing with Lapped Transforms*, Norwood, MA: Artech House, 1992.

2. J. Princen, A. Johnson, and A. Bradley, "Subband Transform Coding Using Filter Bank Designs Based on Time Domain Aliasing Cancellation," in *Proc. IEEE Int. Conf. Acoust., Speech, and Signal Proc.*, (Dallas, TX), April 1987, pp. 2161–2164.

3. C. Herley, J. Kovačević, K. Ramchandran, and M. Vetterli, "Tilings of the Time-Frequency Plane: Construction of Arbitrary Orthogonal Bases and Fast Tiling Algorithms," *IEEE Trans. Signal Proc., Special Issue on Wavelets and Signal Processing*, vol. 41, December 1993, pp. 3341–3359.

4. Y. Meyer, *Wavelets: Algorithms and Applications*, Philadelphia, PA: SIAM, 1993. Translated and revised by R. D. Ryan.

5. I. Daubechies, *Ten Lectures on Wavelets*, Philadelphia, PA: SIAM, 1992.

6. C. Todd, G. Davidson, M. Davis, L. Fielder, B. Link, and S. Vernon, "AC-3: Flexible Preceptual Coding for Audio Transmission and Storage," in *Convention of the AES*, (Amsterdam, Holland), February 1994.

7. J. Kovačević, D. LeGall, and M. Vetterli, "Image Coding with Windowed Modulated Filter Banks," in *Proc. IEEE Int. Conf. Acoust., Speech, and Signal Proc.*, (Glasgow, UK), May 1989, pp. 1949–1952.

8. A. Johnson, J. Princen, and M. Chan, "Frequency Scalable Video Coding using the MDCT," in *Proc. IEEE Int. Conf. Accoust., Speech, and Signal Proc.*, (Adelaide, Australia), 1994.

9. R. Coifman and Y. Meyer, "Remarques sur l'analyse de Fourier à Fenêtre," *C. R. Acad. Sci. Paris*, vol. I, 1991, pp. 259–261.

10. P. Auscher, G. Weiss, and M. V. Wickerhauser, "Local Sine and Cosine Bases of Coifman and Meyer and the Construction of Smooth Wavelets," in *Wavelets: A Tutorial in Theory and Applications* (C. Chui, ed.), pp. 237–256, San Diego: Academic Press, 1992.

11. N. Jacobson, *Basic Algebra I*, W. H. Freeman, 1985.

12. X.-G. Xia, and B. W. Suter, "A Family of Two Dimensional Nonseparable Malvar Wavelets," 1994. Preprint.

13. R. Bernardini and J. Kovačević, "Local Orthogonal Bases II: Window Design," *Multidim. Syst. and Sign. Proc., Special Issue on Wavelets and multiresolution Signal Processing*, July 1996. Invited paper.

14. J. Serre, *Linear Representations of Finite Groups*, New York: Springer-Verlag, 1977.

15. L. Debnath and P. Mikusinski, *Introduction to Hilbert Spaces with Applications*, New York, NY: Academic Press, 1990.

16. N. Akhiezer and I. Glazman, *Theory of Linear Operators in Hilbert Spaces*, vol. 1, Frederick Ungar Publisher, 1966.

References

Multidimensional Systems and Signal Processing, 7, 371–399 (1996)
© 1996 Kluwer Academic Publishers, Boston. Manufactured in The Netherlands.

Local Orthogonal Bases II: Window Design

RICCARDO BERNARDINI   bernardi@lcavsun1.epfl.ch
EPFL, CH-1015 Lausanne, Switzerland

JELENA KOVAČEVIĆ   jelena@research.bell-labs.com
Bell Laboratories, Murray Hill, NJ 07974, USA

Abstract. In the first part of this work we presented a technique to find a local orthogonal basis for a given vector space. The concept of a local orthogonal basis can be seen as an extension of the one-dimensional local cosine basis used, for example, in audio processing. Here the problem of window design is discussed, with a particular emphasis on the two-dimensional case, both in continuous and discrete time.

1. Introduction

In the first part of this work [1] we presented a general theory for the construction of local orthogonal bases. The scenario is the following: we have a set X and a vector space $\mathcal{V}_X$ of functions defined on X and we want to find a basis $b_{i,j}$, $i, j = 1, 2, \ldots$, such that each function $b_{i,j}$ has finite support B_i.

Such a goal is achieved by decomposing $\mathcal{V}_X$ as a direct sum of vector spaces with support B_i. In order to obtain the vector space relative to B_i, each B_i is partitioned into a disjoint union of opportune $C_{i,j}$ and the vector space on B_i is defined as the direct sum of some suitable vector space with support $C_{i,j}$.

With each $C_{i,j}$ is associated a group Γ of involutions of $C_{i,j}$, that is, a group of invertible functions mapping $C_{i,j}$ in itself such that $u(u(x)) = x$, for each $x \in C_{i,j}$. Each involution $u \in \Gamma$ acts on the functions defined on $C_{i,j}$ in a natural way $f_u(x) \stackrel{\Delta}{=} f(u(x))$ and to each involution u we associate a weight $s(u) = \pm 1$ such that $s(uv) = s(u)s(v)$, that is, s is a *representation* of Γ.

The vector space with support $C_{i,j}$ is therefore defined as the vector space of functions f that can be written as $f = wS$, where w is a suitable "window function" with support $C_{i,j}$ and S is a function with support $C_{i,j}$ such that, for each $u \in \Gamma$, $S(u(x)) = s(u)S(x)$, $x \in C_{i,j}$, where $s(u)$ is the weight assigned to $u \in \Gamma$.

The conditions that group Γ, weights $s(u)$ and window w must satisfy in order to obtain an orthogonal decomposition of $\mathcal{V}_X$ are presented in the first part of this work [1], where the reader can also find all the details pertaining to notation.

It is worth remembering, as a result from the first part, that window w must satisfy the condition of power-complementarity:

$$\sum_{u \in \Gamma} w^2(u(x)) = 1 \tag{1}$$

115

and that the projection from $\mathcal{V}_X$ to the space associated to $C_{i,j}$ is

$$P(\Gamma, s, w)f \stackrel{\Delta}{=} \sum_{u \in \Gamma} s(u) w w_u f_u. \tag{2}$$

In [1], we presented also an algorithmic procedure to obtain such a basis. First of all, the sets B_i have to be chosen, then each set B_i is decomposed into a disjoint union of suitable $C_{i,k}$, such that for each B_j and $C_{i,k}$ either $C_{i,k} \subset B_j$ or $C_{i,k} \cap B_j = \emptyset$. For each $C_{i,k}$ a suitable set of symmetries with the respective weights are chosen and from such symmetries a set of constraints for the window are obtained. A window satisfying the constraints can then be chosen to obtain the corresponding orthogonal decomposition of $\mathcal{V}_X$.

In this second part we will present a technique to design the window. We will concentrate, in particular, on the two-dimensional, case, both in continuous and discrete time. Our goal will be to design a window for these cases. We will proceed as follows:

Section 2 will discuss the problem in discrete time, in particular in two dimensions. In Section 2.1 the sets B_i in which X is decomposed will be chosen. For the sake of simplicity, each B_i will be defined as a translation of a set B on a regular two-dimensional lattice. Next, in Section 2.2 each B_i will be partitioned into a union of $C_{i,j}$. Because of the "regularity" of the choice of B_i, the definition of $C_{i,j}$ is simplified. As a third step, in Section 2.3, the structure of vector spaces $\mathcal{V}_{C_{i,j}}$ will be defined, that is, for each $C_{i,j}$ the corresponding involution group $\Gamma_{i,j}$ and representation $s_{i,j}$ will be chosen. As a consequence, a set of constraints for the window w will be obtained in Section 2.4.

It is worth noting that the method is quite general and can be applied to a variety of other cases. Indeed, an example of a window with hexagonal support will be given in Section 2.5.

An important property that can be required from a filter bank, especially in the case of image compression, is the *polyphase normalization*, that is, a constant signal should give a nonnull output only in the low-frequency channel [2]. In Section 2.6 we elaborate on the conditions that must be met in order to achieve polyphase normalization.

In the general case, the problem of designing the window can be described as the problem of finding a window that satisfies the constraints we described and is optimal in some sense.

The procedure of window design will differ, depending on the optimality criterion. In Section 2.7 we show how to design the window when the optimality criterion has a specific form. Finally, in Section 2.8 an example of window design is given for the two-dimensional discrete-time case.

Section 3 will follow a similar path, but in continuous time.

2. Designing the Window in Discrete Time

2.1. Choice of Sets B_i

To make the discussion clear, we concentrate here on the two-dimensional discrete-time case. We will choose sets B_i as translations of an "original" set B. For the sake of simplicity let B be a rectangle $[0 \ldots N_x - 1] \times [0 \ldots N_y - 1]$ and define the other sets B_i as

$$B_i = B + \mathbf{Nn} \tag{3}$$

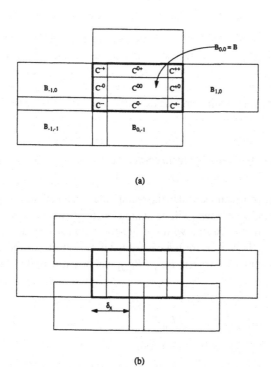

(a)

(b)

Figure 1. Sets $B_{i,j}$ and their decomposition into $C_{i,j}^{s_1,s_2}$. (a) $\delta_x = 0$. (b) $\delta_x > 0$.

with $\mathbf{n} \in \mathbf{Z}^2$ and matrix $\mathbf{N}$ equal to

$$\mathbf{N} = \begin{bmatrix} N_x - \Delta_x & \delta_x \\ 0 & N_y - \Delta_y \end{bmatrix} \tag{4}$$

with $N_x - \Delta_x, \delta_x, N_y - \Delta_y > 0$ and $N_x - \Delta_x > \delta_x$. The points $\mathbf{Nn}$, with $\mathbf{n} \in \mathbf{Z}^2$, are regularly spaced on the plane and form a *lattice* of basis $\mathbf{N}$ [3] that will be denoted as $\Lambda \mathbf{N}$. It is worth noting that, given a certain lattice, it is always possible to choose a basis $\mathbf{N}$ of the form (4) [3]. In the following it will be convenient to change the notation and to call $B_{i,j}$ the set B translated by $\mathbf{N}_{i,j}^T$ (instead of our notation B_i). Often set $B_{0,0}$ will be denoted as B for notational convenience.

We will assume that N_x, N_y, Δ_x and Δ_y are even and that

$$N_x \leq 2\Delta_x,$$
$$N_y \leq 2\Delta_y. \tag{5}$$

Sets $B_{i,j}$ are depicted in Figures 1(a) and 1(b), respectively, when $\delta_x = 0$, or $\delta_x > 0$. It is clear that hypothesis (5) implies that a rectangle has a nonempty intersection only with its "immediate neighbors". Note that the "classical" case of two-dimensional local cosine bases obtained from the one-dimensional ones is with $\Delta_x = N_x/2$ and $\Delta_y = N_y/2$.

We will first concentrate on the rectangular case and we will examine the nonrectangular one later. To start, we have to decompose each $B_{i,j}$ according to Section 3 of the first part.

2.2. Decomposition of B_i in the Rectangular Case

Looking at Figure 1(a) one can see that in the rectangular case each set $B_{i,j}$ can be partitioned in nine different zones. Such zones will be denoted as $C_{i,j}^{s_1 s_2}$, where s_1, s_2 $in\{+, 0, -\}$. Note again that this is not in accordance with our general notation (we decomposed X into B_i and each B_i into $C_{i,j}$). However, for the case at hand, it is more convenient to simplify the notation. The intersection of two sets B_{i_1,j_1}, B_{i_2,j_2} can be expressed as the union of certain zones, for example

$$\begin{aligned} B_{0,0} \cap B_{-1,0} &= C_{0,0}^{--} \cup C_{0,0}^{-0} \cup C_{0,0}^{-+} \\ &= C_{-1,0}^{+-} \cup C_{-1,0}^{+0} \cup C_{-1,0}^{++}. \end{aligned} \tag{6}$$

It is worth noting that the same set can sometimes be interpreted in different ways. For example, from (6) it is obvious that set $C_{0,0}^{--}$, i.e., the lower left zone of B, can be interpreted also as $C_{-1,0}^{+-}$, that is the lower right zone of $B_{-1,0}$.

Next step will be to decide the structure of the vector space associated with each zone. It is important to note that the vector space $cV_{C_{i,j}^{st}}$ associated with the zone s, t of set $B_{i,j}$ can be obtained by translating by $N[i, j]^T$ the functions belonging to $V_{C_{0,0}^{st}}$. Such a fact implies that we can limit ourselves to studying only the case of B because of the regularity of the decomposition.

2.3. Selection of the Vector Spaces

Zone C^{00}, the central one, has no intersection with other zones, therefore its vector space has to be $\chi_{C^{00}}V_X$.

Looking at zone C^{+0} in Figure 1 one can see that it intersects only with the zone $C_{1,0}^{-0}$ of the set $B_{1,0}$, therefore, it must be $V_{C^{+0}} \oplus V_{C_{1,0}^{-0}} = \chi_{C^{+0}}V_X$, that is, the sum of the two vector spaces relative to C^{+0} and $C_{1,0}^{-0}$ must yield the vector space of the functions belonging to V_X and having support C^{+0}. This can be achieved by

- choosing as our symmetry group $\{\mathcal{I}, u_y\}$, where u_y is the symmetry with respect to the central vertical axis of C^{+0},

- imposing that the window w_+, relative to zone C^{+0}, be symmetric to the window w_-, relative to zone C^{-0} and

- choosing two different signs for $s_+(u_y)$ and $s_-(u_y)$.

These are governed by Property 2 in [1]. Of course, both windows will have to satisfy the condition of power-complementarity, as given by (27) in [1]. It is worth noting that the condition that N_x, N_y, Δ_x and Δ_y be even implies that no point is symmetric to itself. This is not a fundamental condition and although it could be easily removed, removing it would complicate the study of the window without giving more information. It is clear that an analog reasoning holds for zones C^{0-} and C^{0+}.

The four remaining zones $(C^{--}, C^{-+}, C^{+-}, C^{++})$ in Figure 1(a) intersect each other and the condition of completeness becomes

$$V_{C^{--}} \oplus V_{C^{-+}} \oplus V_{C^{+-}} \oplus V_{C^{++}} = \chi_{C^{--}} V_x \tag{7}$$

Goal (7) can be fulfilled by making the following choices:

- The group relative to each set will be $\Gamma_4 \triangleq \{\mathcal{I}, u_x, u_y, u_{xy}\}$ where u_x and u_y are the symmetries with respect the horizontal and vertical axes, respectively, and u_{xy} is the combination of the two, i.e., the symmetry around the center of the zone. Note that $\Gamma_4 = \langle \Gamma_x \Gamma_y \rangle$, where $\Gamma_x = \{\mathcal{I}, u_x\}$ and $\Gamma_y = \{\mathcal{I}, u_y\}$.

- Because of decomposition of group Γ reported in the previous point, one can make the following association projection-zone:

$$
\begin{aligned}
C^{++} &\Rightarrow P_x P_y, \\
C^{-+} &\Rightarrow P_x^* P_y, \\
C^{+-} &\Rightarrow P_x P_y^*, \\
C^{--} &\Rightarrow P_x^* P_y^*,
\end{aligned}
\tag{8}
$$

where P_x is a projection made with only a horizontal symmetry, P_y is made with only a vertical symmetry and P_x^* and P_y^* are, respectively, the complementary projections of P_x and P_y as given in Definition 2 [1]. It is clear that the sum of the space relative to C^{++} with the one relative to C^{+-} gives a space whose projection is $P_x P_y + P_x P_y^* = P_x(P_y + P_y^*) = P_x$, i.e., a projection relative to a horizontal symmetry. Note that the windows relative to P_y and P_x must satisfy the constraint specified in Property 3 of the first part.

By imposing the structure (8), the window w relative to set B can be decomposed into the product of two windows, w_h, w_v, relative, respectively, to the projection with the horizontal

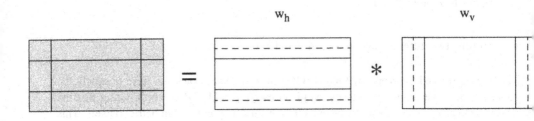

Figure 2. The decomposition of the window w corresponding to the set B as a product of the horizontal and vertical windows, w_h and w_v, respectively.

and vertical symmetries, as in Figure 2 (recall also Property 3 from [1]). In the following, window w_h will be called the *horizontal* window and window w_v the *vertical window*. As can be seen in Figure 2, window w_h has a central rectangle ($C^{-0} \cup C^{00} \cup C^{+0}$) in which $w_h = 1$ and two other rectangles in which it must verify the condition of power complementarity. The portion of w_h relative to $C^{-+} \cup C^{0+} \cup C^{++}$ will be called the *upper tail* and the portion relative to $C^{--} \cup C^{0-} \cup C^{+-}$ will be called the *lower tail*.

2.4. *Constraints on the Window*

As was shown in the previous sections, the window w cannot be freely chosen. More precisely, it must satisfy the power complementarity condition (1) in order for $P(\Gamma, s, w)$ to be a projection and it must satisfy some symmetry constraints in order to grant the orthogonality of the spaces with the same support.

It is worth noting that, in general, the constraints have a "local" nature; for example, in the one-dimensional case, the right tail of window w_0 is required to be symmetric to the left tail of w_1 and the left tail of w_0 has to be symmetric to the right one of w_{-1}, but there is no relation between the left and the right tails of w_0. Such a restriction comes when we want to use the same window for each interval. In this case, the left tail of w_1, symmetric to the right one of w_0, is equal to the left tail of w_0 and this makes a new constraint between the values of w_0.

In the two-dimensional case, if we drop the requirement that each $B_{i,j}$ uses the same window, the portion of window on $C_{0,0}^{0+}$ has constraints only with $C_{0,1}^{0-}$, but if we want to use everywhere the same window, we have a constraint between $C_{0,0}^{0+}$ and $C_{0,0}^{0-}$.

In this section, by using the vector space structures chosen in the previous section and

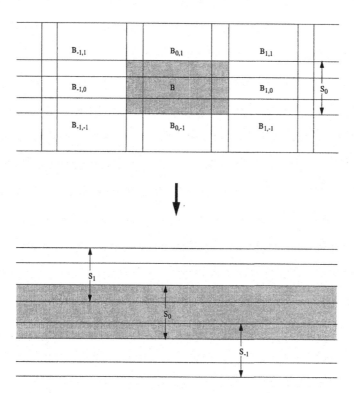

Figure 3. The strip generated by the vector spaces of row 0 (rectangular case).

by imposing the use of the same window for every set, we will figure out the constraints required and a form of the window, as a function of some free parameters.

It is useful to start with the following: sum all spaces relative to row zero (that is, the spaces relative to sets $B_{i,0}$, with $i \in \mathbf{Z}$). Since the vertical symmetries cancel one other, one obtains a vector space whose support is a horizontal "strip" S_0 with two zones of horizontal symmetry (Figure 3). Note also that, by construction, the window relative to such a strip is periodic with period $N_x - \Delta_x$. Note that by summing the vector spaces of row j one obtains the corresponding strip S_j. When the vector spaces of two adjacent strips are summed the symmetries on the common transition zone compensate one another and a vector space having a "larger" strip as support is obtained. In this way one can fill all the plane $\mathbf{Z}^2$.

Using such a fact we can construct the projection on set B in the following way:

- Project from $\mathcal{V}_X$ to $\mathcal{V}_{S_0}$. This require a projection of type (2) with an opportune "horizontal" window w_H having a transition zone of height Δ_y and periodic along x of period $N_x - \Delta_x$, as in Figure 5(a).

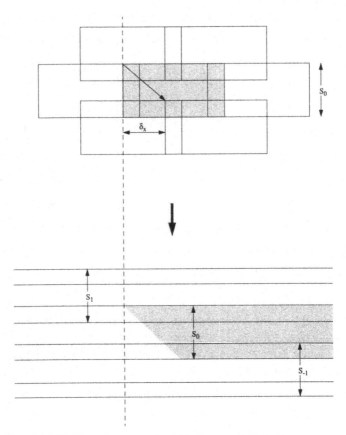

Figure 4. The strip generated by the vector spaces of row 0 in the nonrectangular case.

- Project from cV_{S_0} to $\chi_B V_{S_0}$. This only requires to "truncate" the function outside B. Note that the sequence of these two steps is equivalent to projecting functions of V_X with a horizontal window w_h (which is one period of w_H) having width N_x and whose "corner" $[0 \ldots \Delta_x - 1] \times [0 \ldots \Delta_y - 1]$ is equal to the "corner" $[N_x - \Delta_x + 1 \ldots N_x] \times [0 \ldots \Delta_y - 1]$, as in Figure 5(b).

- Project from $\chi_B V_{S_0}$ to V_B using a projection of type (2) with a vertical window w_v with transition zones of width Δ_x.

The resulting window will be the product of the window w_h with the window w_v. Note that in order to do the third step w_h must be vertically symmetric in the left and right borders and w_v must be horizontally symmetric.

This approach is interesting because it can be used also when $\delta_x \neq 0$. Indeed, if translations of V are made on a nonrectangular lattice, as in Figure 1(b), when all the spaces of a given row are summed one again obtains a vector space defined in a "strip", but in this case the strip S_j is translated by δ_x with respect to S_{j-1}. This imposes that, in order to have orthogonality and completeness, the upper tail must be obtained from the lower one via a horizontal symmetry and a translation of $-\delta_x$ (Figure 4).

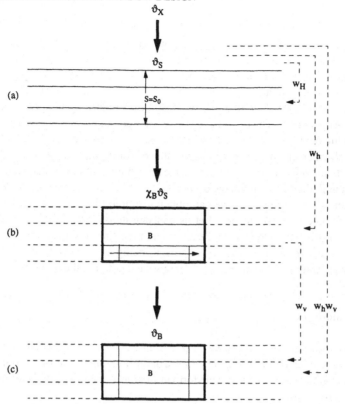

Figure 5. Construction of the projection on the set B (a) Projection from cV_X to V_{S_0}. (b) Projection from V_{S_0} to $\chi_B V_{S_0}$. (c) Projection from $\chi_B V_{S_0}$ to V_B.

In essence, the horizontal window w_H must satisfy the following constraints:

- The upper tail must be symmetric to the lower one with, eventually, a translation of $-\delta_x$.

- Both the upper and the lower tails must be periodic with period $N_x - \Delta_x$.

- The two corners of upper and lower tails must be vertically symmetric. Note that the symmetry of the upper tail, if $\delta_x \neq 0$, induces an "image" symmetry on the lower one, due to the first point. If $\delta_x = 0$ the "normal" symmetry and the "image" ones coincide (Figure 4).

- The lower tail must satisfy the condition of power-complementarity with respect to the horizontal symmetry.

From the first constraint one obtains that the upper tail is uniquely determined by the lower one and, moreover, from the fourth it is clear that the upper half of the lower tail determines the lower half of the same tail. Therefore, the set of free points is a subset of $[0 \ldots \Delta_y/2 - 1] \times [N_x - \Delta_x \ldots \Delta_y]$. However, not every point of such a rectangle is free because of the vertical symmetry induced by the second point. It is worth noting that if the "main" corner and the "image" one are distinct and with nonempty intersection, to explicitly show the equality constraints in a closed form can be very complex. A very simple algorithm to figure out all equality constraints induced by the previous points can be found in Appendix B. Once we have the set of free points we can assign to each one of them a value between 0 and 1 with which we can construct a valid horizontal window. Note that each value can be thought of as the sine of some angle between 0 and $\pi/2$.

For the vertical window one can use the same reasoning, a little simpler, since there is no vertical shift and, therefore, no image corner.

By using these constraints now we can obtain a simple form expressing the window as a function of some free parameters. Let us concentrate on the horizontal window. Each sample of the horizontal window can be considered the sine of some angle between 0 and $\pi/2$. With $\sin(\mathbf{n})$, where $\mathbf{n} = [n_1, n_2, \ldots, n_M]^T$ being a real vector, we will denote the vector $[\sin(n_1), \sin(n_2), \ldots, \sin(n_M)]^T$. Using such a notation the horizontal window can be expressed as

$$w_h = \sin(\alpha_h) \tag{9}$$

where each component of α_h is between 0 and $\pi/2$. Note that in (9) w_h is considered a column vector obtained by ordering the samples of the horizontal window in some opportune way. The constraint on w_h can be transformed in constraints on α_h. Recall that we have two types of constraints:

- Equality constraints, that is, $w_h[i] = w_h[j]$ for some i, j.

- Power complementarity constraints, that is, $w_h^2[i] + w_h^2[j] = 1$ for some i, j.

Such constraints, when translated into α_h, become

$$\begin{aligned} \alpha_h[i] &= \alpha_h[j], \\ \alpha_h[i] &= \pi/2 - \alpha_h[j]. \end{aligned} \tag{10}$$

Moreover, for some points it must be $w_h[i] = 1$, that is, $\alpha_h[i] = \pi/2$. A convenient way of expressing α_h can be

$$\alpha_h = C_h f_h + K_h \tag{11}$$

where C_h is a matrix with entries $-1, 0$ or 1, f_h is a "free" vector and K_h is a constant vector with entries 0 or $\pi/2$. Using (11) one can write the horizontal window as

$$w_h = \sin\left(C_h f_h + K_h\right), \tag{12}$$

the vertical one as

$$w_h = \sin\left(C_v f_v + K_v\right), \tag{13}$$

and the global window vector as

$$w = w_h w_v = \sin(\mathbf{C}_h \mathbf{f}_h + \mathbf{K}_h) \odot \sin(\mathbf{C}_v \mathbf{f}_v + \mathbf{K}_v) \tag{14}$$

where $\odot$ denotes the pointwise product. Right hand of (14) can be written in a more interesting form by using the trigonometric identity $\sin(\alpha)\sin(\beta) = 1/2[\cos(\alpha - \beta) - \cos(\alpha + \beta)]$:

$$\frac{1}{2}[\cos(\mathbf{C}_h \mathbf{f}_h + \mathbf{K}_h - \mathbf{C}_v \mathbf{f}_v - \mathbf{K}_v) - \cos(\mathbf{C}_h \mathbf{f}_h + \mathbf{K}_h + \mathbf{C}_v \mathbf{f}_v + \mathbf{K}_v)] \tag{15}$$

that can be expressed in matrix form

$$\underbrace{\frac{1}{2}[\,-\mathbf{I}\ \ \mathbf{I}\,]}_{\mathbf{A}}\cos\left(\underbrace{\begin{bmatrix} \mathbf{C}_h & \mathbf{C}_v \\ \mathbf{C}_h & -\mathbf{C}_v \end{bmatrix}}_{\mathbf{C}}\underbrace{\begin{bmatrix} \mathbf{f}_h \\ \mathbf{f}_v \end{bmatrix}}_{\mathbf{f}} + \underbrace{\begin{bmatrix} \mathbf{K}_h \\ \mathbf{K}_v \end{bmatrix}}_{\mathbf{K}}\right). \tag{16}$$

By using (16), equation (14) can be rewritten as

$$w = \mathbf{A}\cos(\mathbf{Cf} + \mathbf{K}) \tag{17}$$

where $\mathbf{f}$ is a vector belonging to $[0, \pi/2]^M$, with M the number of the components of $\mathbf{f}$.

2.5. Filters with Hexagonal Support

Set B is not required to be a rectangle; for example, it could be hexagon-like, as depicted in Figure 6(a). The shape in Figure 6(a) is obtained by "expanding" the edges of a hexagon in order to create six transition zones where adjacent sets can intersect.

It is clear from the figure that the symmetry will be around each edge and the conditions of symmetry and power-complementarity are simple to obtain. It is worth noting that the theory developed for the rectangular case applies also to the hexagonal one. This implies that all its results are valid and that the window design method explained in Section 2.7 can also be used with filters having this shape.

Note that the choice of a hexagonal support implies a nonrectangular sampling lattice when these projections are interpreted like filter banks.

2.6. Polyphase Normalization

A property that can be required from a local cosine filter bank is that of *polyphase normalization*. Such a property states that a constant signal should give a nonnull output only in the low-frequency channel [2]. This is desirable, for example, in image compression because the lack of polyphase normalization causes visible artifacts [2].

In effect, with the vector space point of view, the concept of low-pass channel is partially lost and all the channels look alike; therefore, the condition of polyphase normalization

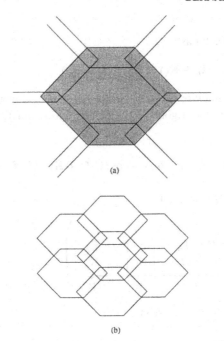

(a)

(b)

Figure 6. Hexagonal support and covering of $\mathbf{Z}^2$ obtained with it.

can be restated as the request that a constant signal must give nonnull output in only one channel.

Recall that in [1] we showed that the general structure of an orthonormal basis of $\mathcal{V}_B$ is $\mathcal{WKG}$, where $\mathbf{G}$ is a unitary matrix and contains free parameters. Then, it is clear that the projection of a constant must be a vector of basis $\mathbf{G}$.

In order to solve such a problem, consider B partitioned as $B = \cup_{s_1,s_2} C^{s_1,s_2}$. As it was seen in Section 2.7 of the first part, each function of $\mathcal{V}_B$ is uniquely determined by its values on $\tilde{B}$, where $\tilde{B}$ denotes the set of representative of B, that is, each function belonging to $\mathcal{V}_B$ is uniquely determined by its values on $\tilde{B}$ (the same holds for $\tilde{C}_j$). Let $x \in \tilde{C}_j$ and write the condition which states that $g(x)$ must be equal to the projection of a constant signal evaluated in x (use (2))

$$g(x) = \sum_{u \in \Gamma} s(u)w(x)w(u(x)). \tag{18}$$

Equation (18) imposes certain constraints on $g(x)$ (depending on the structure of group Γ) in order to be able to find a window satisfying (18).

If $\Gamma = \{\mathcal{I}\}$, then equation (18) implies that $g(x) = 1$, or, more generally, $g(x) = \alpha$ for some $\alpha \in \mathbf{R}$. That is, in zones of B without symmetry, like C^{00}, in order to achieve polyphase normalization, function g must be constant.

If $\Gamma = \{\mathcal{I}, u\}$, then the sum in (18) has two terms. By remembering that $w^2(x) + w^2(u(x)) = 1$ one can write, without any loss of generality, $w(x) = \sin \alpha$, $w(u(x)) = \cos \alpha$.

With such an assumption (18) becomes

$$g(x) = \sin^2 \alpha + s(u) \sin \alpha \cos \alpha. \tag{19}$$

Using trigonometric identities (19) can be transformed into

$$g(x) = -\frac{1}{\sqrt{2}} \cos\left(2\alpha + s(u)\frac{\pi}{4}\right) + \frac{1}{2}. \tag{20}$$

If $\Gamma = \{\mathcal{I}, u_h, u_v, u_{hv}\}$, then the values of $w(u(x))$, $u \in \Gamma$ can be expressed as $w(x) = \sin\alpha \sin\beta$, $w(u_h(x)) = \sin\alpha \cos\beta$, $w(u_v(x)) = \cos\alpha \sin\beta$ and $w(u_{hv}(x)) = \cos\alpha \cos\beta$. Then, (18) can be rewritten as

$$g(x) = \left(-\frac{1}{\sqrt{2}} \cos\left(2\alpha + s(u)\frac{\pi}{4}\right) + \frac{1}{2}\right)\left(-\frac{1}{\sqrt{2}} \cos\left(2\beta + s(u)\frac{\pi}{4}\right) + \frac{1}{2}\right). \tag{21}$$

It is worth noting that the values of w on $\tilde{B}$ are not free, but they must satisfy certain constraints. Let us return for a moment to the two-dimensional discrete-time case we discussed in Section 2.2 (there the notation is slightly different, instead of C_i we use C^{s_1,s_2}). For example, the lower half of C^{0+} must be equal to the upper one of C^{0-}. This implies, via (20), the equality of the corresponding values of $g(x)$. Similar constraints can be obtained also for the values of $g(x)$ on the corners C^{++}, C^{+-}, C^{-+} and C^{--}.

In conclusion, the condition of polyphase normalization does not depend only on the window, but both on the basis **G** and the window. Moreover, not every vector g admits a window satisfying the polyphase normalization, but there are some constraints on g deriving from the corresponding constraints of w.

An interesting observation is that the problem of finding a window such that the system is polyphase normalized can be seen upside down, that is, we first choose the window, then (18) will give a vector of the basis meeting the normalization. The other vectors of the basis can be subsequently chosen.

2.7. Optimal Window Design

Let us start the section on window design by stating a simple window function property that descends from power complementarity.

PROPERTY 1 *Let w be a window on set B. Then $\sum_{n \in B} w^2[n] = k$, where k is a constant depending on both B and the symmetries.*

The proof of Property 1 is given in Appendix A. Property 1 is important because it states that the window lies on the surface of a multidimensional sphere and this gives a sort of "rigidity" to window w. More precisely, if w is a window, then αw, with $|\alpha| \neq 1$ cannot be a window. By using Property 1 with the Parseval's theorem the following corollary is immediate:

COROLLARY 1 *Let $W(\lambda)$ the Fourier transform of the window w from Property 1. Then,* *by denoting with $\int \int_{-\pi}^{\pi} \ldots d\lambda$ the integral on the square $[-\pi \ldots \pi] \times [-\pi \ldots \pi]$,*

$$\int \int_{-\pi}^{\pi} |W(\lambda)|^2 \, d\lambda = k. \tag{22}$$

If we want to design the window in some optimal sense, we have to select a figure of merit for the window. A general cost function is the following:

$$\sigma(w) = \sum_i \int \int_{-\pi}^{\pi} \phi_i(\lambda)|W(\lambda) - g_i(\lambda)|^2 \, d\lambda \tag{23}$$

where ϕ_i are nonnegative weight functions and $g_i(\lambda)$ are functions describing the wanted behavior of $W(\lambda)$, the Fourier transform of window ww. Cost function (23) can describe several common cost functions. For example, by choosing $\phi_i(\lambda) = \chi_R$, (23) becomes the power in the stop-band

$$\int_R |W(\lambda)|^2 \, d\lambda, \tag{24}$$

or if we want that $W(\lambda)$ is close to $g_1(\lambda)$ in R_1 and close to $g_2(\lambda)$ in R_2, with the region outside $R_1 \cup R_2$ as a transition zone, we can use

$$\int_{R_1} |W(\lambda) - g_1(\lambda)|^2 \, d\lambda + \int_{R_2} |W(\lambda) - g_2(\lambda)|^2 \, d\lambda. \tag{25}$$

Cost function (23) can be used also to measure the frequential "spreading" as defined in [4]

$$\sigma_f = \frac{\int \int_{-\pi}^{\pi} |\lambda|^2 |W(\lambda)|^2 \, d\lambda}{\int \int_{\pi}^{\pi} |W(\lambda)|^2 \, d\lambda} \tag{26}$$

since in (26) the denominator is constant, because of Corollary 1.

However, cost function (23) has a problem because of the window "rigidity" seen before. For example, if we are required to minimize the distance of $W(\lambda)$ from constant 1 we would obtain too "difficult" a problem, since if $W(\lambda) \approx 1$, then it should be $\int \int_{-\pi}^{\pi} |W(\lambda)|^2 \, d\lambda \approx 4\pi^2$, but, Corollary 1 implies that $\int \int_{-\pi}^{\pi} |W(\lambda)|^2$ is a constant that can be different from $4\pi^2$.

For this reason we will allow an arbitrary scale factor by multiplying g_i in (23) by a scalar α_i

$$\sigma(w) = \sum_i \int_{R_i} \phi_i(\lambda)|W(\lambda) - \alpha_i g_i(\lambda)|^2 \, d\lambda. \tag{27}$$

In this case, $g_i(\lambda)$ gives an indication of the "shape" of W and the effective "scale" is left undetermined. Of course, (27) will have to be minimized with respect to both α_i and $w[\mathbf{n}]$.

For the moment, we will elaborate on the single term of sum in (27) to bring it in a more suitable form. Window $w[\mathbf{n}]$ has support B and one can write

$$W(\lambda) = \sum_{\mathbf{n} \in B} w[\mathbf{n}] e^{-j\lambda^T \mathbf{n}}. \tag{28}$$

By substituting (28) in (27), and rewriting it, one obtains

$$\sigma(w) = \begin{bmatrix} \mathbf{W}^T & | & \alpha_i \end{bmatrix} \begin{bmatrix} \mathbf{R}_{ww} & \mathbf{R}_{wg} \\ \hline \mathbf{R}_{wg}^T & R_{gg} \end{bmatrix} \begin{bmatrix} w \\ \hline \alpha_i \end{bmatrix} \tag{29}$$

where $\mathbf{R}_{ww}$, $\mathbf{R}_{wg}$ and R_{gg} are, respectively, a square matrix, a column vector and a scalar whose values are given explicitly in the following. Vector $\mathbf{w}$ contains the values of window $w[\mathbf{n}]$ ordered in some way, for example $w[\mathbf{n}] = w_{d[\mathbf{n}]}$, where $d[\mathbf{n}]$ denotes ordering. This derivation is given in Appendix A. With such an ordering, the matrices in (29) are

$$R_{gg} = \int \int_{-\pi}^{\pi} \phi_i(\lambda) |g_i(\lambda)|^2 \, d\lambda,$$

$$\mathbf{R}_{ww}[d[\mathbf{n}], d[\mathbf{m}]] = \int \int_{-\pi}^{\pi} \phi_i(\lambda) e^{-j\lambda^T(\mathbf{n}-\mathbf{m})} \, d\lambda,$$

$$\mathbf{R}_{wg}[d[\mathbf{n}]] = \int \int_{-\pi}^{\pi} \phi_i(\lambda) e^{-j\lambda^T \mathbf{n}} g_i^*(\lambda) e\lambda. \tag{30}$$

The cost function in (29) is a quadratic form and, as well known, we can use matrix $(\mathbf{R}^T + \mathbf{R})/2$ instead of $\mathbf{R}$. The entries in (30) become

$$R_{gg} = \int \int_{-\pi}^{\pi} \phi_i(\lambda) |g_i(\lambda)|^2 \, d\lambda$$

$$\mathbf{R}_{ww}[d[\mathbf{n}], d[\mathbf{m}]] = \int \int_{-\pi}^{\pi} \phi_i(\lambda) \cos(\lambda^T(\mathbf{n} - \mathbf{m})) \, d\lambda$$

$$\mathbf{R}_{wg}[d[\mathbf{n}]] = \int \int_{-\pi}^{\pi} \phi_i(\lambda) \cos(\lambda^T \mathbf{n}) \Re(g_i(\lambda)) \, d\lambda$$

$$- \int \int_{-\pi}^{\pi} \phi_i(\lambda) \sin(\lambda^T \mathbf{n}) \Im(g_i(\lambda)) \, d\lambda \tag{31}$$

where $\Re(x)$ and $\Im(x)$ mean, respectively, the real and the imaginary part of x. It is interesting to note that the values of $\mathbf{R}_{ww}$ and $\mathbf{R}_{wg}$ can be obtained from the Fourier coefficients of $\phi_i(\lambda)$, $\phi_i(\lambda)\Re(g_i(\lambda))$ and $\phi_i(\lambda)\Im(g_i(\lambda))$. Observe that in the important case of $g_i(\lambda) = 0$ (used, for example, in the case of power in stop-band or frequental spreading) $R_{gg} = 0$ and $\mathbf{R}_{wg} = 0$ and matrix in (29) can be simplified.

So far, we put a single term of sum (27) in matrix product form (30). Now, suppose to have two terms in sum (23), with analog reasoning we can write each one as

$$\sigma_1(w) = \begin{bmatrix} \mathbf{w}^T & | & \alpha_1 & | & \alpha_2 \end{bmatrix} \begin{bmatrix} \mathbf{R}_{1,ww} & \mathbf{R}_{1,wg} & 0 \\ \mathbf{R}_{1,wg}^T & R_{1,gg} & 0 \\ \hline 0 & 0 & 0 \end{bmatrix} \begin{bmatrix} \mathbf{w} \\ \alpha_a \\ \alpha_2 \end{bmatrix},$$

$$\sigma_2(w) \;=\; \left[\; \mathbf{w}^T \,\middle|\, \alpha_1 \,\middle|\, \alpha_2 \;\right] \left[\begin{array}{c|c|c} \mathbf{R}_{2,ww} & \mathbf{0} & \mathbf{R}_{2,wg} \\ \hline \mathbf{0} & 0 & 0 \\ \hline \mathbf{R}_{2,wg}^T & 0 & R_{2,gg} \end{array}\right] \left[\begin{array}{c} \mathbf{w} \\ \alpha_1 \\ \alpha_2 \end{array}\right]. \tag{32}$$

By calling $\mathbf{R}_1$ and $\mathbf{R}_2$ the two square matrices in (32) and denoting with $\hat{\mathbf{w}}$ the "extended" window vector $[\mathbf{w}^T, \alpha_1, \alpha_2]^T$, we can sum the two partial cost functions to obtain the total cost function

$$\sigma(w) = \sigma_1(w) + \sigma_2(w) = \hat{\mathbf{w}}^T (\mathbf{R}_1 + \mathbf{R}_2)\hat{\mathbf{w}}. \tag{33}$$

It is clear that (27) can always be written as $\hat{\mathbf{w}}^T \mathbf{R} \hat{b w}$, where $\mathbf{R}$ is an opportune matrix and vector $\hat{\mathbf{w}}$ accounts both for the values of the windows and of the free multipliers α_i.

2.8. Design Example

As an example, a two-dimensional window has been designed by using as cost function the power in the band outside of the square $[-1/16\pi \ldots 1/16\pi] \times [-1/16\pi \ldots 1/16\pi]$. The window has support 16×16 and is designed for a rectangular subsampling 8×8.

The results of the optimization are depicted in the right column of Figure 7, while in the left column of the same figure we give a bidimensional separable window obtained by multiplying two monodimensional ones. The contour level of the window is given in part (b), while its Fourier transform magnitude, both as contour level and three-dimensional perspective plot, is depicted in parts (d) and (f). Parts (a), (c) and (e) give the same information for the separable window.

The separable nature of the window obtained from the monodimensional one is clearly evident, especially in subfigure (c), while the window designed in a true bidimensional way is clearly nonseparable.

The fact that the presented procedure allows us to design nonseparable windows is important since it gives us much more freedom in choosing the window. It is clear that the optimal nonseparable window for a given application cannot be worse than the separable one, because the latter is a particular case of the former. (Here, obviously, "nonseparable" means "not necessarily separable.")

An example of application in which nonseparable windows are desirable can be found in [5], [6] where it is suggested that nonseparable Malvar wavelets (that can be interpreted as dual to the local cosine bases) can be useful in simulations of turbulent flows because separable wavelets introduce artifacts at corners of rectangular regions.

3. Designing the Window in Continuous Time

In this section we will show how one can proceed in order to obtain a window for the two-dimensional continuous-time case, characterized by an arbitrary degree of smoothness. We will concentrate ourselves on the horizontal window w_h since the construction of the vertical one can be carried out in a similar way.

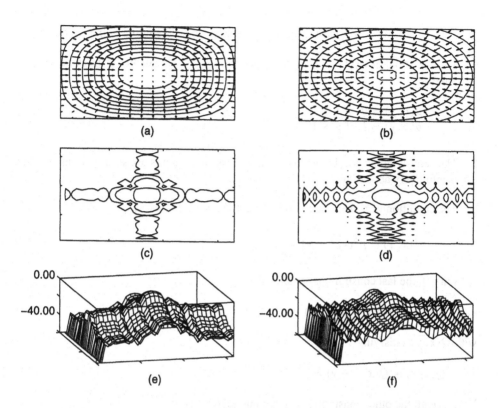

Figure 7. Example of window design with passband $[-1/16\pi \ldots 1/16\pi] \times [-1/16\pi \ldots 1/16\pi]$. (a), (c), (e) Contour plot, magnitude Fourier transform contour plot, magnitude Fourier transform of the separable window. (b), (d), (f) Contour plot, magnitude Fourier transform contour plot, magnitude Fourier transform of the nonseparable window. In (c), (e), (d) and (f) the frequency axes are between $-\pi$ and π and the magnitude is in dB. Contour lines are regularly spaced at 20 dB from one another.

First, let us normalize the problem: by changing the scale we can impose the window support to be $[0, 1 + \Delta_x] \times [0, 2]$. The motivation for such a support will be clear in the following.

By mimicking the approach in [7] and since it simplifies the construction, we will construct the window w_h as the product of a semi-window with itself time-reversed around the line $y = 1$:

$$w_h(x, y) \overset{\Delta}{=} p(x, y)p(x, 2 - y). \tag{34}$$

On the semi-window the following constraints will be imposed:

1. The semi-window will be arbitrarily smooth.

2. Semi-window $p(x, y)$ has to be periodic along the x-axis:

$$p(x, y) = p(x + 1, y). \tag{35}$$

131

3. The semi-window will have the same "symmetry zones" required of the window.[1]

4. The constraint of $w_h(x, y)$ being zero outside $0 \leq y \leq 2$ is satisfied if the semi-window satisfies

$$
\begin{aligned}
p(x, y) &= 0 \ \text{ if } y < 0 \\
p(x, y) &= 1 \ \text{ if } y > 1.
\end{aligned}
\tag{36}
$$

5. The semi-window will be monotone in y. This constraint is not strictly necessary, but is intuitively appealing.

6. The condition of power normalization has to be verified, that is,

$$
p^2(x, y) + p^2(x, 1 - y) = 1.
\tag{37}
$$

As in [7], the last constraint is satisfied if

$$
p(x, y) = \sin(\pi Q(x, y)/2)
\tag{38}
$$

with $Q(x, y)$ satisfying

$$
Q(x, y) + Q(x, 1 - y) = 1.
\tag{39}
$$

Note that all the other constraints of p are inherited by Q.

Let us study the monotonicity condition. Such a constraint is equivalent to requiring that the partial derivative of $Q(x, y)$, taken with respect to y, be nonnegative. Call $q(x, y)$ such a derivative, then we have

$$
Q(x, y) = \int_0^y q(x, t) \, dt.
\tag{40}
$$

The constraints of continuity, periodicity and symmetry are inherited by $q(x, y)$, while the "boundary conditions" (36) become conditions on the mean value of $q(x, y)$, for each x, that is

$$
\int_0^1 q(x, t) \, dt = 1.
\tag{41}
$$

Moreover, smoothness constraints imply that

$$
q(x, y) = 0 \quad y \notin [0, 1].
\tag{42}
$$

Condition (39) can be rewritten, by using (40), as

$$
\int_0^y q(x, t) \, dt + \int_0^{1-y} q(x, t) \, dt = 1.
\tag{43}
$$

By a change of variable in the second integral, (43) becomes

$$\int_0^y q(x,t)\,dt + \int_y^1 q(x,1-t)\,dt = 1 \tag{44}$$

Equating (41) with (44) one obtains

$$\int_0^y q(x,t)\,dt + \int_y^1 q(x,1-t)\,dt = \int_0^y q(x,t)\,dt + \int_y^1 q(x,t)\,dt \tag{45}$$

that is,

$$\int_y^1 q(x,1-t)\,dt = \int_y^1 q(x,t)\,dt \tag{46}$$

Equation (46) implies that, if $q(x,t)$ is continuous,

$$q(x,1-t) = q(x,t). \tag{47}$$

That is, the power normalization is equivalent to the symmetry of $q(x,y)$ around $y = 1/2$.

Within the inherited requirements there is the periodicity of $q(x,y)$. This implies that $q(x,y)$ can be expressed as a Fourier series

$$q(x,y) = \sum_{n\in\mathbf{Z}} e^{j2\pi nx} g(n,y). \tag{48}$$

Let us see how the constraints on q map themselves in constraints on g.

1. The requirement of smoothness is verified if $g(n,y)$ is smooth enough in y and is nonnull for all but a finite number of n. Note that these conditions are stronger than the smoothness of q since, for example, we can have q continuous even with an infinite number of $g(n,y)$ not equal to zero.

2. The periodicity requirement is automatically satisfied.

3. The symmetry requirement is a linear constraint and becomes a linear constraint on the components $g(n,y)$. We will not further elaborate on this point and in the example below we will handle it in a simplified way. Since we will certainly have symmetry around $x = 0$ we can decide to keep only the "cosine part" in series (48). This is stronger than required, but it simplifies the reasonings.

4. In order to obtain the corresponding of the new boundary conditions (41) and (42), express them using (48)

$$\int_0^1 \sum_{n\in\mathbf{Z}} e^{j2\pi nx} g(n,y)\,dy = 1 \qquad \text{if } y > 1$$

$$\sum_{n\in\mathbf{Z}} e^{j2\pi nx} g(n,y) = 0 \qquad \text{if } y \notin [0,1]. \tag{49}$$

Because g(n,y) is zero for all but a a finite number of n, the sum inside the integral in (49) has only a finite number of terms and can be moved outside. Therefore, conditions (49) become

$$\int_0^1 g(n, y)\,dy = \delta_n \tag{50}$$

$$g(n, y) = 0 \qquad \text{if } y \notin [0, 1].$$

5. The nonnegativity of $q(x, y)$ becomes

$$\sum_{n\in\mathbf{Z}} \cos(2\pi nx)g(n, y) \geq 0. \tag{51}$$

Because cosines have even symmetry, in the following we will suppose that $g(n, y) = 0$ for $n < 0$. We will elaborate on (51) in a moment.

6. The symmetry of q around $y = 1/2$ implies the same symmetry for $g(n, y)$.

So far we can easily handle each constraint but the requirement of nonnegativity of $q(x, y)$. In order to simplify it, let us observe that, for each y, (51) can be interpreted as the Fourier transform of the following sequence:

$$\hat{g}(n, y) = \begin{cases} g(n, y) & n \geq 0 \\ g(-n, y) & n < 0 \end{cases} \tag{52}$$

obtained by extending $\hat{g}(n, y)$ by symmetry around the origin. Since sequences obtained by autocorrelation have nonnegative Fourier transforms, one can try to obtain $g(n, y)$ from an autocorrelation. More precisely, choose a function $h(k, y)$ and define

$$g(n, y) = \begin{cases} \sum_k h(k, y)h(k + n, y) & n \geq 0 \\ 0 & n < 0 \end{cases}, \tag{53}$$

that is, $g(n, y)$ is made of the values of the autocorrelation of $h(k, y)$, for $n \geq 0$. Such $g(n, y)$ satisfies (51).

Now we have to map the constraints of g into constraints on h.

1. The requirement of smoothness is inherited from g. Note that if $h(n, y) = 0$ for all but a finite number of n, $g(n, y)$ has only a finite number of nonzero values as well.

2. The constraints (50) become

$$\int_0^1 \sum_{k\in\mathbf{Z}} h(k, y)h(k + n, y)\,dy = \delta_n \tag{54}$$

and

$$\sum_k h(k, y)h(k + n, y) = 0 \qquad \text{if } y \notin [0, 1]. \tag{55}$$

Equation (55) is verified if $h(k, y) = 0$ for $y \notin [0, 1]$. We will elaborate on (54) in a moment.

3. Equation (51) is automatically verified.

4. The symmetry of $g(n, y)$ around $y = 1/2$ is inherited by $h(k, y)$.

To see how (54) can be satisfied observe that the left-hand side can be interpreted as the scalar product between $h(k, y)$ and its translation by n, that is, (55) requires that $h(k, y)$ and its translations make an orthonormal set. It can be proved that such a condition is equivalent to

$$\int_0^1 |H(\omega, y)|^2 \, dy = 1 \tag{56}$$

for each ω. However, it is simpler to require that

$$\int_0^1 h(k, y)h(k + n, y) \, dy = \delta_n. \tag{57}$$

One can obtain $h(k, y)$ satisfying equation (57) by choosing functions $\hat{h}(k, y)$ and orthogonalizing them using, for example, the Gram-Schmidth procedure.

Summarizing, in order to obtain the horizontal window do the following:

- Choose functions $\hat{h}(k, y)$ such that they are

  1. smooth enough in y,
  2. zero when $y \notin [0, 1]$,
  3. symmetric with respect to $y = 1/2$.

- Orthogonalize them using the Gram-Schmidth procedure and obtain $h(k, y)$. Note that all constrains on $\hat{h}(k, y)$ are linear; therefore, $h(k, y)$, being linear combination of $\hat{h}(k, y)$ will satisfy them too.

- Compute the autocorrelation (53) of $h(k, y)$ and obtain $g(n, y)$.

- Use $g(n, y)$ as Fourier coefficients in (48) and find $q(x, y)$.

- Integrate $q(x, y)$ with respect to y to obtain $Q(x, y)$.

- Apply (38) to $Q(x, y)$, resulting in the semi-window $p(x, y)$.

- Finally, $p(x, y)$ is used in (34) to obtain $w_h(x, y)$.

- Apply again this procedure to obtain the vertical window $w_v(x, y)$.

- Find the total window w by

$$w(x, y) = w_h(x, y)w_v(x, y). \tag{58}$$

3.1. Design Example

In this section we will show how, by means of a specific example, to construct a window in the continuous-time case. It is worth observing that our aim is to show how to construct a window that is smooth, that is, we will not have any particular objective such as a given frequency response or stopband attenuation.

To make the example more complete, we will construct a window that can be used for hexagonal sampling. To this end, choose the window structure as shown in Figure 8. It is worth observing that the central symmetry zone for the horizontal window is due to the "image corner" induced by the nonrectangular sampling. Note the peculiar frame chosen in Figure 8; such a frame will make easier writing the constraints for the window $w_h(x, y)$ and the semi-window $p(x, y)$. More precisely, the constraints on the semi-window are:

1. Smoothness.

2. Periodicity:

$$p(x, y) = p(x + 1, y). \qquad (59)$$

3. Symmetry. According to Figure 8, $w_h(x, y)$ and, therefore, $p(x, y)$ have to be symmetric with respect to $x = 0$ and $x = 1/2$, that is

$$
\begin{aligned}
p(x, y) &= p(-x, y) \\
p(x, y) &= p(1 - x, y).
\end{aligned}
\qquad (60)
$$

4. Boundary conditions:

$$
\begin{aligned}
p(x, y) &= 0 \qquad \text{if } y < 0 \\
p(x, y) &= 1 \qquad \text{if } y > 1.
\end{aligned}
\qquad (61)
$$

5. Monotonicity:

$$y_0 \geq y_1 \Rightarrow p(x, y_0) \geq p(x, y_1), \quad \forall x. \qquad (62)$$

6. Power normalization: ·

$$p^2(x, y) + p^2(x, 1 - y) = 1. \qquad (63)$$

Note that condition (60) is automatically fulfilled if we choose to express $p(x, y)$ as a Fourier series with only cosines with period 1. This is because of the particular form of condition (60); in a more general case the second symmetry point cannot be chosen to be $1/2$ if we fix the period length to 1, but in most of cases we can choose the second symmetry point as fraction of 1 and the symmetry constraints can be fulfilled by simply using only a subset of the cosines.

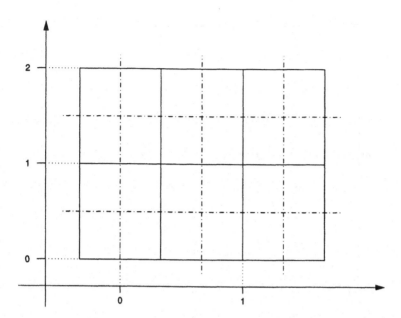

Figure 8. Structure of a continuous-time window suitable for hexagonal sampling. Dotted vertical segments denote the symmetries for the horizontal window w_h; while the horizontal ones denote the symmetries for the vertical window w_v.

As pointed out in the previous section, we need only to choose a function $\hat{h}(k, y)$, with $k \in \mathbf{Z}$ and $y \in [0, 1]$, smooth in y, symmetric with respect to $y = 1/2$ and with support $[0, 1]$. In this case function

$$\hat{h}(k, y) \stackrel{\Delta}{=} \exp\left(\frac{1}{(y - y^2)(k + 1)}\right) \tag{64}$$

has been chosen for the horizontal window and only the first six cosines $(1, \cos(2\pi x), \ldots, \cos(2\pi 5x))$ have been used. It is easy to see that function (64) is very smooth; it is continuous and with all derivatives continuous. This is a classical example used to show the existence of functions with all derivatives continuous and compact support. A plot of (64) for $k = 0$ is given in Figure 9.

Function (64) has been processed with the algorithm presented in Section 3 and the resulting horizontal window is displayed in Figure 10(a) with a three-dimensional plot and in Figure 11 as contour levels. Note in Figure 11 the window symmetry around $x = 0$ and $x = 1/2$.

The vertical window, shown in Figure 10(b), has been constructed, for sake of simplicity, constant in the vertical direction. Such a fact has been accomplished by using only the first cosine, namely the constant 1, by using

$$\hat{h}(k, y) = \left\{ \begin{array}{ll} 1 & \text{if } k = 0 \\ 0 & \text{else.} \end{array} \right\} \tag{65}$$

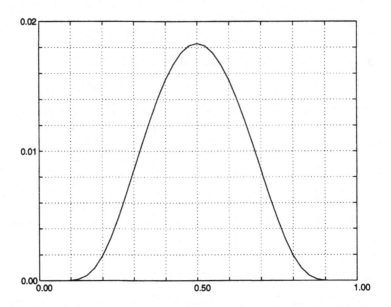

Figure 9. Plot of function $\hat{h}(k, y)$ for $k = 0$ used to construct the horizontal window shown in Figure 10(a).

The total window, resulting by the product of the horizontal and the vertical windows, is shown in Figure 10(c).

4. Conclusions

In this work we discussed window design for the local orthogonal bases presented in [1] and, in particular, we concentrated on the the two-dimensional case both in continuous and discrete time. The technique is quite general and can be adapted to a wide range of applications.

Acknowledgements

We would like to thank anonymous reviewers for their fruitful comments.

A   Proof of properties

Proof of Property 1: Let B be partitioned as $B = \cup_{s_1,s_2} C^{s_1,s_2}$ and let Γ_{s_1,s_2} be the symmetry group relative to C^{s_1,s_2}. Compute the sum of squares of window w as

$$\sum_{s_1,s_2} \sum_{\mathbf{n} \in C^{s_1,s_2}} w^2[\mathbf{n}]. \tag{66}$$

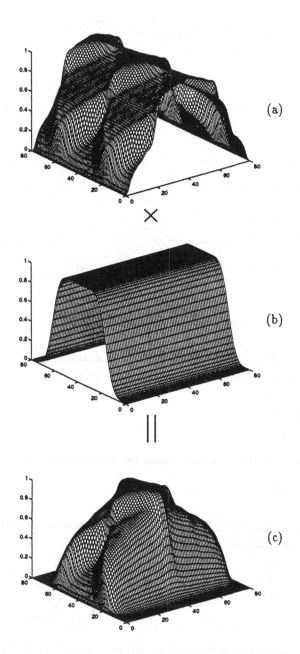

Figure 10. Example of window construction for the continuous-time case. (a) Three-dimensional plot of one period of the horizontal window $w_h(x, y)$. (b) Three-dimensional plot of one period of the vertical window $w_v(x, y)$. (c) Three-dimensional plot of the final window, product of (a) and (b).

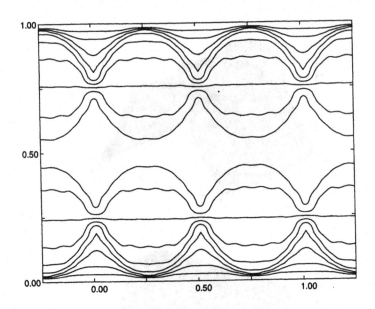

Figure 11. Contour level plot of the horizontal window of Figure 10(a).

In order to show that (66) is a constant, it is sufficient to show that the inner sum is constant. If $\tilde{C}^{s_1,s_2}$ is a set of representatives of the orbits induced on C^{s_1,s_2} by Γ_{s_1,s_2}, one can rewrite the inner sum of (66) as

$$\sum_{\mathbf{n}\in\tilde{C}_{s_1,s_2}}\sum_{\mathbf{m}\in O(\mathbf{n})} w^2[\mathbf{m}], \tag{67}$$

where $O(\mathbf{n})$ is the orbit of $\mathbf{n}$. It is known, from the power normalization, that

$$\sum_{u\in\Gamma_{s_1,s_2}} w^2[u[\mathbf{n}]] = 1. \tag{68}$$

If $\text{Stab}(\mathbf{n}) \neq \{\mathcal{I}\}$, in sum (68) there are some equal terms, more precisely, by remembering that $\text{Stab}(\mathbf{n})$ is a subgroup of Γ_{s_1,s_2}, one can rewrite (68) as

$$\sum_{v\in\text{Stab}(\mathbf{n})}\sum_{u\in\Gamma_{s_1,s_2}/\text{Stab}(\mathbf{n})} w^2[u[v[\mathbf{n}]]] = \sum_{v\in\text{Stab}(\mathbf{n})}\sum_{u\in\Gamma_{s_1,s_2}/\text{Stab}(\mathbf{n})} w^2[u[\mathbf{n}]]$$

$$= |\text{Stab}(\mathbf{n})| \sum_{u\in\Gamma_{s_1,s_2}/\text{Stab}(\mathbf{n})} w^2[u[\mathbf{n}]]. \tag{69}$$

A theorem of group theory [8] states that $u[\mathbf{n}]$, when $u \in \Gamma_{s_1,s_2}/\text{Stab}(\mathbf{n})$, generates all elements of $O(\mathbf{n})$, once and only once. Therefore, by comparing (69) and (68), one can write

$$\sum_{u\in\Gamma_{s_1,s_2}} w^2[u[\mathbf{n}]] = |\text{Stab}(\mathbf{n})| \sum_{\mathbf{m}\in O(\mathbf{n})} w^2[\mathbf{m}] = 1, \tag{70}$$

that is,

$$\sum_{\mathbf{m} \in O(\mathbf{n})} w^2[\mathbf{m}] = 1/|\text{Stab}(\mathbf{n})|. \tag{71}$$

By substituting (71) in (67) one obtains that

$$\sum_{\mathbf{n} \in C^{s_1, s_2}} w^2[\mathbf{n}] = \sum_{\mathbf{n} \in \bar{C}^{s_1, s_2}} 1/|\text{Stab}(\mathbf{n})|, \tag{72}$$

which is a constant.   ∎

Proof of (29):

$$\begin{aligned}
\sigma(w) &= \int\int_{-\pi}^{\pi} \phi_i(\lambda) \left(\sum_{\mathbf{n}} w[\mathbf{n}] e^{-j\lambda^T \mathbf{n}} - \alpha_i g_i(\lambda) \right) \left(\sum_{\mathbf{n}} w[\mathbf{n}] e^{-j\lambda^T \mathbf{n}} - \alpha_i g_i(\lambda) \right)^* d\lambda \\
&= \int\int_{-\pi}^{\pi} \phi_i(\lambda) \sum_{\mathbf{n},\mathbf{m}} w[\mathbf{n}] w[\mathbf{m}] e^{-j\lambda^T(\mathbf{n}-\mathbf{m})} d\lambda \\
&\quad -\alpha_i \int\int_{-\pi}^{\pi} \phi_i(\lambda) \sum_{\mathbf{n}} w[\mathbf{n}] e^{-j\lambda^T \mathbf{n}} g_i^*(\lambda) d\lambda \\
&\quad -\alpha_i \int\int_{-\pi}^{\pi} \phi_i(\lambda) \sum_{\mathbf{n}} w[\mathbf{n}] e^{-j(-\lambda)^T \mathbf{n}} g_i(\lambda) d\lambda \\
&\quad +\alpha_1^2 \int\int_{-\pi}^{\pi} \phi_i(\lambda) |g_i(\lambda)|^2 d\lambda \\
&= \sum_{\mathbf{n},\mathbf{m}} w[\mathbf{n}] w[\mathbf{m}] \int\int_{-\pi}^{\pi} \phi_i(\lambda) e^{-j\lambda^T(\mathbf{n}-\mathbf{m})} d\lambda \\
&\quad -\alpha_i \sum_{\mathbf{n}} w[\mathbf{n}] \int\int_{-\pi}^{\pi} \phi_i(\lambda) e^{-j\lambda^T \mathbf{n}} g_i^*(\lambda) d\lambda \\
&\quad -\alpha_i \sum_{\mathbf{n}} w[\mathbf{n}] \int\int_{-\pi}^{\pi} \phi_i(\lambda) e^{+j\lambda^T \mathbf{n}} g_i(\lambda) d\lambda \\
&\quad +\alpha_i^2 \int\int_{-\pi}^{\pi} \phi_i(\lambda) |g_i(\lambda)|^2 d\lambda \tag{73}
\end{aligned}$$

∎

B An algorithm to find the equality constraints

In Section 2.4 it was pointed out that, in the design of the window for the local orthogonal system, the horizontal window must satisfy some equality constraints. It was noted that, when $\delta_x \neq 0$, the symmetry constraints and the periodicity ones could combine to give

0	1	2	3	4	5	6	7	8	9	10	11	12	13	14	15
a_1	a_2	a_3	a_3	a_2	a_1	*	*	*	*	a_1	a_2	a_3	a_3	a_2	a_1
	b_1	b_2	b_3	b_3	b_2	b_1	*	*	*	*	b_1	b_2	b_3	b_4	

Figure 12. Example of interactions between equality constraints in a horizontal window.

very complex constraints between values of w_h. In this section we will give an algorithm to figure out such constraints.

A first observation is that such constraints do not depend on the vertical position and they are the same for every row of w_h, so we can limit our study to only one row of w_h.

Let us illustrate, with an example, what kind of interactions we have between the two type of constraints. Figure 12 shows the case of a horizontal window with $N_x = 16$, six points of "corner" and $\delta_x = 2$. Let $w_h[n]$ be the window. The first row of Figure 12 gives the values of n. The second row shows the equality constraints due to the symmetry of the corner and to the periodicity. The third row, finally, gives the equality constraints due to the symmetry of the image corner. It is clear form the second row that, for example, $w_h[0] = w_h[5]$, but, from the third row we can see that $w_h[5] = w_h[4]$ and this imply also $w_h[0] = w_h[4]$ and from the second row it follows $w_h[4] = w_h[1]$ that implies $w_h[0] = w_h[1] = w_h[4] = w_h[5]$ and so on.

It is clear that it is very difficult to give the equality constraints in a closed form that depends on N_x, Δ_x and δ_x. However, there exists a simple algorithm to compute them.

Indeed, we want to know, for each couple of integer n_1, n_2 between 0 and N_x, if there is a chain of integers $m_0, m_1, \ldots, m_\ell$, such that $w_h[n_1] = w_h[m_0] = \cdots = w_h[m_\ell] = w_h[n_1]$, where each equality descends directly from the symmetry constraints of the corner or of the image corner. This is the problem of finding the *transitive closure* of the relation "a and b are such that $w_h[a] = w_h[b]$ because of the corner or the image corner symmetry" [9].

The easiest way to compute the transitive closure of a relation between M elements is to construct an $M \times M$ matrix $\mathbf{R}$, such that $\mathbf{R}_{ij} = 1$ if and only if the element i and the element j are in such relation and zero otherwise. The transitive closure of $\mathbf{R}$ can be computed as [9]:

$$\mathbf{I} + \mathbf{R} + \mathbf{R}^2 + \cdots + \mathbf{R}^M \tag{74}$$

where the product and the sum used in (74) are, respectively, the boolean AND and the boolean OR.

Expression (74) can be slow to compute (although it is not too slow for windows of normal size and current computers) and a more efficient algorithm exists, see [10].

Notes

1. Although it could be possible to formally write this constraint, for the moment we will use this description for sake of simplicity. In the specific example below we will be more explicit.

References

1. R. Bernardini and J. Kovačević, "Local Orthogonal Bases I: Construction," *Multidim. Syst. and Sign. Proc., special. issue on Wavelets and Multiresolution Signal Processing*, July 1996. Invited paper.

2. H. Malvar, *Signal Processing with Lapped Transforms*, Norwood, MA: Artech House, 1992.

3. M. Newman, *Integral Matrices*, New York, NY: Academic Press, 1972.

4. S. Venkataraman and B. Levy, "Nonseparable Orthogonal Linear Phase Perfect Reconstruction Filter Banks and Their Application to Image Compression," in *Proc. ICIP '94*, vol. 3, IEEE, Nov. 1994.

5. X.-G. Xia and B. W. Suter, "A Family of Two Dimensional Nonseparable Malvar Wavelets," 1994. Preprint.

6. X.-G. Xia and B. W. Suter, "Construction of Malvar Wavelets on Hexagons," 1994, Preprint.

7. P. Auscher, G. Weiss, and M. V. Wickerhauser, "Local Sine and Cosine Bases and Coifman and Meyer and the Construction of Smooth Wavelets," in *Wavelets: A Tutorial in Theory and Applications* (C. Chui, ed.), pp. 237–256, San Diego: Academic Press, 1992.

8. N. Jacobson, *Basic Algebra I*, W. H. Freeman, 1985.

9. G. Birkhoff and S. MacLane, *Introduction to Graph Theory*, Longman, 1953.

10. S. Warshall, "A Theorem on Boolean Matrices," *JACM*, vol. 9, no. 1, 1962, pp. 11–12.